职业教育教材

分析检验技术专业英语

Professional English for Analysis and Inspection Technology

陈 佳　隋 欣　编

化学工业出版社

·北京·

内 容 简 介

本书分为化学基础，健康、安全和环境（HSE），分析化学（化学分析和仪器分析）和分析检验技术4部分，介绍了分析检验技术专业英语的基本词汇和实用表达，配合类型丰富的习题、相关基础语法和泛读材料，结合附录，可供读者方便地学习掌握专业知识和技能中常用的词汇、构词法、缩写和符号、音标等。

全书内容全面实用，合理编排难度梯度，由浅入深，易于阅读和理解。对课文中出现的生词、术语、操作用语等都进行了注释，习题部分也强调了学生对专业术语及操作用语的掌握，拓展阅读部分与化学化工专业知识和技能密切相关、且具有一定的趣味性，并配有相应的中文翻译，便于学生阅读及理解。

本书适合高等职业学校和中等职业学校分析检验技术专业师生教学使用，也可供生物与化工、食品药品与粮食、资源环境与安全、能源动力与材料、医药卫生等相关专业的专业英语教学和查阅使用，还可作为相关企业人员的参考书。

图书在版编目（CIP）数据

分析检验技术专业英语/陈佳，隋欣编．—北京：化学工业出版社，2022.8
ISBN 978-7-122-41629-2

Ⅰ.①分⋯ Ⅱ.①陈⋯②隋⋯ Ⅲ.①化学分析-英语-教材②化工产品-质量检验-英语-教材 Ⅳ.①O65②TQ075

中国版本图书馆 CIP 数据核字（2022）第 099944 号

责任编辑：提　岩　旷英姿　　　　　　　文字编辑：曹　敏
责任校对：杜杏然　　　　　　　　　　　　装帧设计：王晓宇

出版发行：化学工业出版社（北京市东城区青年湖南街 13 号　邮政编码 100011）
印　　装：河北鑫兆源印刷有限公司
787mm×1092mm　1/16　印张 13　字数 323 千字　2023 年 1 月北京第 1 版第 1 次印刷

购书咨询：010-64518888　　　　　　　　　售后服务：010-64518899
网　　址：http://www.cip.com.cn
凡购买本书，如有缺损质量问题，本社销售中心负责调换。

定　价：38.00 元　　　　　　　　　　　　　　　　　　　版权所有　违者必究

前言
PREFACE

为落实构建新时代中国特色职业教育体系，国家对职业教育的支持力度不断加大。同时，企业对高素质化学化工行业人才的需求也在不断提高，结合世界技能大赛等高水平赛事，专业英语的学习变得越来越重要。

本书的选材与设计以基础性、时代性和实用性相结合为原则，力求难度梯度与中高职学生的专业知识和英语水平相匹配，内容由浅入深，符合教学规律，聚焦化学和分析检验技术专业的基础内容——基础化学和分析化学的实用词汇和常用表达方式，增加健康、安全和环境（HSE）内容，与现代化工理念相融合，拓展分析检验技术的应用领域。

本书主要特点如下：（1）每课由中文相关专业知识的热身开始，建立专业知识和英语语言学习的联结，帮助学习者建立自己的中英文知识结构；（2）结合世界技能大赛（化学实验室技术赛项）对专业英语的基本要求，聚焦化学、分析化学和分析检验技术的基础知识，参考权威原版教材的基础知识部分，针对性和专业性更强；（3）精选课文内容，侧重基础词汇、实用语法和常用表达，建立与学生专业知识和英语水平相匹配的难度梯度，供学习者和教师选择学习与教学的内容；（4）丰富习题类型，学以致用，巩固复习；（5）深入浅出地穿插引入实用语法和泛读内容，供学生个性化选择用于提高和拓宽视野；（6）附录部分收录了化学化工和分析检验工作中常用的英语词头和词尾、前后缀、分析化学常用词汇［含基础词汇、化学试剂、分析化学实验室常见仪器（附图）、基本操作］、化学常见英语缩写和符号、国际音标（附音频）、元素周期表，图文并茂，可作为简易实用的专业词汇手册进行查阅和学习。

本书由上海信息技术学校陈佳、隋欣编写，上海城建职业学院李博主审。其中，Part 1、Part 2、Part 3（Unit 1）、Part 4和附录由陈佳编写，Part 3（Unit 2）由隋欣编写。全书由陈佳统稿。

本书编写过程中参阅了大量国内外书籍和科技文献，在此谨向这些书籍和科技文献的作者致以诚挚的谢意！同时，本书的编写和出版得到了化学工业出版社的大力支持，同样表示衷心的感谢！

由于编者水平所限，书中不足之处在所难免，在此敬请读者指正为盼！

编者
2022年5月

目录
CONTENTS

Part 1　ABC of Chemistry 化学基础

Lesson 1　Chemistry 化学 ··· 002
Lesson 2　Important Ideas about Chemical Change: Making New Substances 化学变化的重点：生成新物质 ·· 005
Lesson 3　Physical and Chemical Properties 物理性质和化学性质 ····················· 008
Lesson 4　Atoms 原子 ··· 011
Lesson 5　Elements, Compounds and Mixtures 单质、化合物和混合物 ··············· 015
Lesson 6　Solids, Liquids and Gases 固体、液体和气体 ································ 019
Lesson 7　Nomenclature of Inorganic Compounds 无机物命名法 ····················· 022
Lesson 8　Acids and Bases 酸和碱 ·· 031
Lesson 9　Temperature 温度 ··· 036
Lesson 10　Thermometer 温度计 ·· 039
Lesson 11　Diffusion 扩散 ·· 042
Lesson 12　Catalysis 催化作用 ·· 044
Lesson 13　Reversible Reactions 可逆反应 ··· 046
Lesson 14　Factors Affecting Reaction Rates 影响反应速率的因素 ·················· 049
Lesson 15　Energy and Chemical Energy 能量与化学能 ······························ 053

Part 2　Health, Safety, Environment (HSE) 健康、安全和环境

Lesson 16　Hazards in Chemical Engineering Laboratories 化学工程实验室中的危险 ······ 057
Lesson 17　Environment Protection 环境保护 ·· 062
Lesson 18　Green Chemistry 绿色化学 ··· 067

Part 3　Analytical Chemistry 分析化学

Lesson 19　Analytical Chemistry and Chemical Analysis 分析化学和化学分析 ······ 072
Lesson 20　The Function of Analytical Chemistry 分析化学的应用 ·················· 077

Unit 1　Chemical Analysis 化学分析

Lesson 21　Titration 滴定法 ·· 084
Lesson 22　The Types of Titration 滴定法的类型 ··· 087
Lesson 23　Use of Burettes 滴定管的使用 ·· 090
Lesson 24　Acid-Base Titrations 酸碱滴定法 ··· 094
Lesson 25　Complexometric Titration: EDTA Titration Procedure 络合滴定法：EDTA
　　　　　　络合滴定的过程 ·· 097
Lesson 26　Precipitation Titrations 沉淀滴定法 ··· 101
Lesson 27　Redox Titrations 氧化还原滴定法 ·· 105

Unit 2　Instrumental Analysis 仪器分析

Lesson 28　Gas Chromatography 气相色谱法 ·· 109
Lesson 29　Atomic Absorption Spectrometry 原子吸收光谱法 ···································· 112
Lesson 30　Liquid Chromatography 液相色谱法 ·· 115
Lesson 31　Ultraviolet and Visible Spectrophotometer 紫外-可见分光光度计 ·············· 119
Lesson 32　Infrared Spectrometer 红外光谱仪 ·· 123

Part 4　Analysis and Inspection Technology 分析检验技术

Lesson 33　Steps in Analysis and Inspection 分析检验的流程 ···································· 127
Lesson 34　Sample Preparation 试样的制备 ·· 131
Lesson 35　Organic Compound Identification Using Infrared Spectroscopy 运用红外光谱法
　　　　　　鉴别有机物 ··· 135
Lesson 36　Determination of Sulfur Dioxide in Ambient Air 环境空气中二氧化硫的
　　　　　　测定 ·· 139
Lesson 37　Determination of Water Hardness 水硬度的测定 ····································· 145
Lesson 38　Analysis of Trace Amounts of Organic Materials in Soil and Sediments 土壤与
　　　　　　沉积物中微量有机物的分析 ·· 150
Lesson 39　Determination of Protein in Foodstuffs by the Kjeldahl Method 食品中蛋白质
　　　　　　的测定（凯氏定氮法）··· 154
Lesson 40　Determination of Soluble Sugar in Cereals and Beans Using Iodometry 谷物和豆类
　　　　　　中可溶性糖的测定（碘量法）··· 159
Lesson 41　Pharmacopoeia and Analysis of Aspirin 药典简介和阿司匹林的分析

| | 检验 | 164 |

Lesson 42　Introduction to Chinese Medicine Analysis and Inspection 走进中药的分析检验 ········ 169

Lesson 43　Analysis of the Bisphenol A 双酚A的分析检验 ·········· 173

Appendix 附录

Appendix Ⅰ　Head and End of Common English Words 英语常用词头和词尾 ············ 177
Appendix Ⅱ　Common Chemical Prefix and Suffix 化学专业英语词汇常用前后缀 ······ 180
Appendix Ⅲ　Common Chemical English Words 化学专业常用词汇 ·········· 183
　Foundation 基础知识 ············ 183
　Quantitative Analysis 定量分析 ············ 183
　Reagent 化学试剂 ············ 186
　Common Laboratory Instruments 化学实验室常见仪器 ············ 188
　Manipulation 基本操作 ············ 194
Appendix Ⅳ　Common Chemical Abbreviation and Symbols 化学常见英语缩写与符号 ········ 195
Appendix Ⅴ　The International Phonetic Alphabet 国际音标 ············ 199
Appendix Ⅵ　Periodic Table of the Elements 元素周期表 ············ 200

References 参考文献

Part 1

ABC of Chemistry

化学基础

Lesson 1
Chemistry 化学

Warming up

化学： 是一门以实验为基础的自然科学，主要在原子、分子层面上研究物质的组成、结构、性质及其变化规律的科学，也是一门关于如何创造新物质的科学。

世界由物质组成，主要存在着化学变化和物理变化两种变化形式。作为沟通微观与宏观物质世界的重要桥梁，化学是人类认识和改造物质世界的主要方法和手段之一。从开始用火的原始社会，到使用各种人造物质的现代社会，人类都在享用化学成果。门捷列夫提出的化学元素周期表大大促进了化学的发展。如今很多人称化学为"中心科学"，因为化学为部分科学学科的核心，如材料科学、纳米科技、生物化学等。它是一门历史悠久而又富有活力的学科，与人类进步和社会发展的关系非常密切，它的成就是社会文明的重要标志。

化学变化： 产生了新物质的变化。如：燃烧、钢铁生锈、食物腐烂、粮食酿酒、动植物呼吸、光合作用……

化学性质： 是物质在化学变化中表现出来的性质。如所属物质类别的化学通性：酸性、碱性、氧化性、还原性、热稳定性及一些其他特性。

有机化合物： 通常指含碳元素的化合物。但一些简单的含碳化合物，如一氧化碳、二氧化碳、碳酸盐、碳化物、氰化物等除外。除含碳元素外，绝大多数有机化合物分子中含有氢元素，有些还含氧、氮、卤素、硫和磷等元素。因此，有机化合物是指碳氢化合物及其衍生物，简称为有机物。

现代化学下有五个二级学科：无机化学、有机化学、物理化学、分析化学与高分子化学。

Text

There are different kinds of materials in our universe. Each material has its own characteristics, which is called its properties.

The two sciences, chemistry and physics, are important for the study of materials. Physics is concerned with the general properties, energy and physical changes. By contrast, chemistry is concerned with chemical properties and chemical changes. In chemical changes, materials are transformed into different materials. For example, nitrogen and hydrogen can be combined to ammonia. It is called chemical reactions.

Chemistry is very important in the use of materials. It relates to so many areas of human daily life. Chemists work in different fields of chemistry. Biochemists are interested in chemical processes in living plants and animals. Analytical chemists find ways to separate and identify chemical substances. Organic chemists study substances which contain carbon and hydrogen. Inorganic chemists study most of the other elements.

New Words

chemistry [ˈkemɪstrɪ] n. 化学
material [məˈtɪərɪəl] n. 材料，原料，素材
universe [ˈjuːnɪvɜːs] n. 宇宙，经验领域
characteristic [ˌkærəktəˈrɪstɪk] n. 特性，特征
property [ˈprɒpətɪ] n. 性质，性能
physics [ˈfɪzɪks] n. 物理学
transform [trænsˈfɔːm] v. 改变，转变，转化
nitrogen [ˈnaɪtrədʒən] n. 氮，氮气
hydrogen [haɪdrədʒən] n. 氢，氢气
ammonia [əˈməʊnɪə] n. 氨
combine [kəmˈbaɪn] v. （使）联合，（使）结合
living [ˈlɪvɪŋ] adj. 活的
analytical [ˌænəˈlɪtɪk(ə)l] adj. 分析的，解析的
separate [ˈseprət] adj. 分开的，分离的；v. 分开，隔离
identify [aɪˈdentɪfaɪ] vt. 识别，鉴别
organic [ɔːˈɡænɪk] adj. 有机的，器官的
inorganic [ˌɪnɔːˈɡænɪk] adj. 无机的
contain [kənˈteɪn] vt. 包含，容纳
carbon [ˈkɑːbən] n. 碳
element [ˈelɪmənt] n. 元素，要素

Expressions and Technical Terms

be concerned with 与……有关，涉及……
physical change 物理变化
chemical property 化学性质
chemical change 化学变化
by contrast 和……比起来，对照
be transformed into 被转变成……
chemical reaction 化学反应
be combined to 化合成……，结合成……
relate to 涉及，与……有关
daily life 日常生活
chemical process 化工过程

Exercises

A. Translate the following into English.

1. 物理变化　　2. 化学变化　　3. 物理性质　　4. 化学性质

5. 化学反应　　6. 化学过程　　7. 有机化学家　　8. 无机化学家

9. 分析化学家　　10. 生物化学家

B. Choose the best answer to each question.

1. What is concerned with the general properties, energy and physical changes? (　　)
(A) Physics　　(B) Chemistry　　(C) Biology

2. What is concerned with chemical properties and chemical changes? (　　)
(A) Physics　　(B) Chemistry　　(C) Biology

3. In chemical changes, materials are transformed into (　　) materials.
(A) same　　(B) different　　(C) strange

4. (　　) find ways to separate and identify chemical substances.
(A) Analytical chemists　　(B) Biochemists　　(C) Organic chemists

5. (　　) study substances which contain carbon and hydrogen.
(A) Analytical chemists　　(B) Biochemists　　(C) Organic chemists

C. Answer the following questions.

1. What is chemistry concerned with?
2. What are analytical chemists concerned with?
3. What do organic chemists study?
4. What do inorganic chemists study?

Lesson 2
Important Ideas about Chemical Change: Making New Substances
化学变化的重点：生成新物质

Warming up

化学变化： 产生了新物质的变化。如：燃烧、钢铁生锈、食物腐烂、粮食酿酒、动植物呼吸、光合作用……，宏观上可以看到各种化学变化都产生了新物质，这是化学变化的特征。

化学变化是指相互接触的分子间发生原子或电子的转换或转移，生成新的分子并伴有能量的变化的过程，其实质是旧键的断裂和新键的生成。化学变化过程中总伴随着物理变化。

分类：（1）从反应物和生成物的种类及数量进行划分，可以把化学变化分为四种基本反应类型：化合反应、分解反应、置换反应和复分解反应。

（2）若从反应中元素化合价的升降变化的角度，可以分为氧化还原反应和非氧化还原反应。其中氧化还原反应又分为氧化反应和还原反应，氧化还原反应的实质是发生了电子的转移或偏离。

（3）若从反应中是否有离子参加的角度看，可分为离子反应和非离子反应。离子反应的本质是某些离子浓度发生改变。

（4）若从反应的能量变化的角度看可分为吸热反应和放热反应。

Text

In a chemical reaction

The starting substances, the reactants, react to give new different substances, the products. The changes which take place in the reaction are usually written as an equation.

reactants → products

It means 'reacts to give' (sometimes the reaction conditions are written over the arrow).

Different kinds of chemical change

1. Decomposition

A single substance is broken down into two or more simpler substances. Most metal carbonates decompose to give the oxide and carbon dioxide.

$$CuCO_3 \longrightarrow CuO + CO_2$$

2. Combination

Two substances (usually elements) react together to make a single new compound. Metal and nonmetal: aluminum and iodine react to give aluminum iodide.

$$2Al + 3I_2 \longrightarrow 2AlI_3$$

3. Displacement

These are reactions in which one element takes the place of another. Both metals and nonmetals can do displacements: $e.g.$, more reactive metals can displace less reactive ones from their solutions.

$$Fe + CuSO_4 \longrightarrow FeSO_4 + Cu$$

New Words

substance ['sʌbstəns] $n.$ 物质，材料，实质，内容
product ['prɒdʌkt] $n.$ 产品，结果，乘积，作品
equation [ɪ'kweɪʒn] $n.$ 方程式，等式，相等，反应式
reactant [rɪ'æktənt] $n.$ 反应物
mean [mi:n] $v.$ 表示……的意思，意思是，打算，产生……结果
arrow ['ærəʊ] $n.$ 矢，箭，箭状物，箭头记号
decompose [ˌdi:kəm'pəʊz] $v.$ 分解，(使)腐烂
decomposition [ˌdi:kɒmpə'zɪʃn] $n.$ 分解，腐烂
metal ['metl] $n.$ 金属，金属元素
nonmetal [ˌnɒn'metəl] $n.$ 非金属（元素）
carbonate ['kɑ:bənət] $n.$ 碳酸盐
combination [ˌkɒmbɪ'neɪʃn] $n.$ 结合，联合体，密码组合
compound ['kɒmpaʊnd] $n.$ 复合物
aluminum [ə'lju:mɪnəm] $n.$ 铝
iodine ['aɪədi:n] $n.$ 碘
displacement [dɪs'pleɪsmənt] $n.$ 取代，置换
solution [sə'lu:ʃn] $n.$ 解决，溶液，答案

Expressions and Technical Terms

chemical change 化学变化
chemical reaction 化学反应
take place 发生
reaction condition 反应条件
break down 分解
aluminum iodide 碘化铝

Exercises

A. Translate the following into English.

1. 化学变化　　　2. 化学反应　　　3. 反应物　　　4. 产物

5. 分解反应　　　　6. 化合反应　　　　7. 置换反应　　　　8. 反应条件

9. 分解　　　　　　10. 碘化铝

B. Decide whether the following statements are true (T) or false (F). Write T for true and F for false in each blank.

(　　) 1. In a chemical reaction, the starting substances are the reactants.

(　　) 2. The changes which take place in the reaction are usually written as an equation (reactants→products).

(　　) 3. Sometimes the reaction conditions are written over the arrow of an equation.

(　　) 4. There is a kind of chemical change.

(　　) 5. In combination, a single substance is broken down into two or more simpler substances.

C. Answer the following questions.

1. What are the important ideas about chemical change?
2. What are the different kinds of chemical change?
3. What is decomposition, combination and displacement?

译　文

化学变化的重要概念：生成新物质

在化学反应里

起始物质（反应物）反应生成新的不同的物质（产物）。反应中发生的变化通常用反应式表示。

反应物──→产物

表示"反应生成"（有时候反应条件写在箭头上方）。

不同类型的化学变化

（1）分解反应

一种物质分解变成两种或两种以上更简单的物质。

大多数金属碳酸盐分解生成氧化物和二氧化碳。

$$CuCO_3 \longrightarrow CuO + CO_2$$

（2）化合反应

两种物质（通常是单质）互相反应，生成一种新的化合物。

金属和非金属——铝和碘反应生成碘化铝。

$$2Al + 3I_2 \longrightarrow 2AlI_3$$

（3）置换反应

反应里一种元素取代另一种元素，金属和非金属都能发生置换反应。如较活泼的金属能把较不活泼的金属从它们的溶液里置换出来。

$$Fe + CuSO_4 \longrightarrow FeSO_4 + Cu$$

Lesson 3
Physical and Chemical Properties
物理性质和化学性质

 Warming up

物理变化： 物质发生变化时没有生成新物质，这种变化叫作物理变化。

物理性质： 不通过化学变化就能表现出来的物质性质。

物理性质属于统计物理学范畴，即物理性质是大量分子所表现出来的性质，不是单个原子或分子所具有的。例如：物质的颜色是大量分子集体所具有的性质，是单个分子所不具有的。通常用观察法和测量法来研究物质的物理性质。

化学变化： 物质发生变化时生成新物质，又叫作化学反应。

化学性质： 物质在发生化学变化时才表现出来的性质叫作化学性质。

物质的化学性质由它的结构决定，而物质的结构又可以通过它的化学性质反映出来。物质的用途由它的性质决定。分子是保持物质化学性质的最小粒子，如：馒头遇到固体碘、碘溶液、碘蒸气都会变成蓝色。氧气是分子，而氧气具有的性质氧原子并没有。

任何物质就是通过其千差万别的化学性质与化学变化，才区别于其他物质。

Text

Different kinds of matter have different physical and chemical properties. The properties of a substance are its characteristics. We know one substance from another by their physical and chemical properties. In a physical change, the composition of a substance is not changed. For example, ice can be changed into water. This is a physical change because the composition of a substance is not changed. In a chemical change, the composition of a substance is changed. One or more new substances are formed.

Iron rusts in moist air. When iron rusts, it reacts with the oxygen from the air. A new substance is formed. It is called iron oxide. It has other different properties. Wood will burn if it heated in air. When wood burns, it reacts with the oxygen from the air. New substances are formed. They are carbon dioxide and water. Carbon dioxide and water have different properties. Heat is given off if the combustion of any fuel takes place. The above two cases are chemical changes.

New Words

matter ['mætə(r)] *n.* 事件，问题，重要性，物质
substance ['sʌbstəns] *n.* 物质，材料，实质，内容
composition [kɒmpə'zɪʃ(ə)n] *n.* 成分，组成
rust [rʌst] *n.* 铁锈；*vt.* （使）生锈
moist [mɔɪst] *adj.* 潮湿的；*n.* 潮湿
form [fɔːm] *v.* 形成，构成，组织，塑造
burn [bɜːn] *v.* 使用某物为燃料，烧毁，烧坏，烧伤
heat [hiːt] *v.* 加热，变热，（使）变暖
react [rɪ'ækt] *v.* 反应，使发生相互作用，使起化学反应
combustion [kəm'bʌstʃən] *n.* 燃烧，烧毁，氧化
case [keɪs] *n.* 案例，情形

Expressions and Technical Terms

physical property 物理性质
chemical property 化学性质
physical change 物理变化
chemical change 化学变化
react with… 和……反应
carbon dioxide 二氧化碳
take place 发生

Exercises

A. Translate the following into English.

1. 化学变化　　　　2. 物理变化　　　　3. 化学性质　　　　4. 物理性质

5. 和……反应　　　6. 氧气　　　　　　7. 二氧化碳　　　　8. 燃烧

B. Choose the best answer to each question.

1. When ice is changed into water, it occurs (　　)
（A）a chemical change.
（B）a physical change.
（C）both a chemical and physical change.

2. When iron rusts in moist air, we can say that (　　)
（A）a physical change occurs.
（B）iron reacts with the oxygen from the air.
（C）no other new substances are formed.

3. We know one substance from another by their (　　) properties.
（A）physical and chemical　　（B）physical　　（C）chemical

4. When wood burns, we can say that (　　)
(A) a physical change occurs.
(B) wood reacts with the oxygen from the air.
(C) no other new substances are formed.

C. Decide whether the following statements are true (T) or false (F). Write T for true and F for false in each blank.
(　) 1. The properties of a substance are its characteristics.
(　) 2. In a physical change, the composition of a substance is not changed.
(　) 3. In a chemical change, the composition of a substance is changed.
(　) 4. Heat is not given off if the combustion of any fuel takes place.
(　) 5. Carbon dioxide and water have different properties.

Lesson 4
Atoms 原子

Warming up

原子：指化学反应不可再分的最小微粒。

原子由原子核和绕核运动的电子组成。原子直径的数量级大约是 10^{-10}m，质量主要集中在质子和中子上。

注意："原子是构成物质的最小粒子"是不对的，原子又可以分为原子核与核外电子，原子核又由质子和中子组成，而质子数正是区分各种不同元素的依据。质子和中子还可以继续再分，所以原子不是构成物质的最小粒子，但原子是化学反应中的最小粒子。

化学元素：具有相同的核电荷数（核内质子数）的一类原子的总称。目前，已知的元素有 118 种。原子是一种元素能保持其化学性质的最小单位。

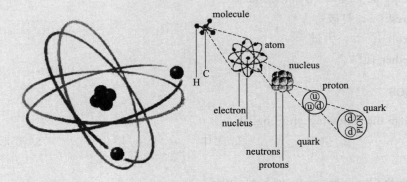

质量数：由于质子与中子的质量相近且远大于电子，所以用原子的质子和中子数量的总和定义原子量，称为质量数。

就算是最重的原子，化学家也很难直接对其进行称量，所以它们通常使用另外一个单位摩尔。

摩尔：对于任意一种元素，一摩尔总是含有同样数量的原子，约为 6.022×10^{23} 个。

Text

Atoms are all around us. They are the bricks of which everything is made. Many millions of atoms are contained in just one grain of salt，but despite their small size they are very important. The way an object behaves depends on what kinds of atoms are in it and how they act.

For instance, you know that most solid objects melt if they get hot enough. Why is this? It is the effect of the heat on the object's atoms. When they are hot, they move faster.

Usually the atoms in an object hold together and give the object its shape. But if the object grows hot, its atoms move so fast that they break the force that usually holds them together. They move out of their usual places so that the object loses its shape. Then we say that the object is melting.

New Words and Expression

atom ['ætəm] n. 原子
brick [brɪk] n. 砖，砖块
grain [greɪn] n. 细粒，颗粒
despite [dɪ'spaɪt] prep. 不管，尽管，不论
object ['ɒbdʒɪkt] n. 物体
behave [bɪ'heɪv] v. 举动，举止，行为
melt [melt] v. （使）融化，（使）熔化
effect [ɪ'fekt] n. 结果，效果
hold [həʊld] n. 把握，控制，掌握；v. 保持，支持
shape [ʃeɪp] n. 外形，形状，形态
break [breɪk] v. 打破
force [fɔːs] n. 力，力量
hold together 使结合

Exercises

A. Translate the following into English.
1. 原子　　　2. 例如　　　3. 固体　　　4. 熔化　　　5. 变形

B. Choose the best answer to each question.
1. One grain of salt contains (　　)
　(A) many millions of atoms.
　(B) several heated atoms.
　(C) one million atoms.
2. The way an object behaves depends on the (　　)
　(A) kinds of atoms in it and how they act.
　(B) number of atoms in it.
3. Atoms in an object move (　　)
　(A) at all times.
　(B) only when the object is heated.
　(C) whenever they grow hot.
4. Heating an object will affect (　　)

（A）the speed of its atoms.
（B）the shape of its atoms.

5. An object hold its shape because its atoms （   ）
（A）usually hold together.
（B）move very fast.
（C）are very hot.

C. Cloze.

The atoms in an object _____ together and give the object its _____. But if the object grows hot，its atoms move so _____ that they _____ the force that usually holds them together. They move out of their usual places so that the object _____ its shape. Then we say that the object is _____ .

Reading Material 阅读材料

科技英语文体的主要特点（一）

1. 科技人员通常使用一些显得稳重的规范词（formal words）（因为普通词汇往往显得比较随便），这样从词汇方面突出了科技英语正式、庄重的语体特征。下面列出一些普通词语和科技英语中的非技术词。

普通英语词（一般语体）	科技英语中的非技术词（正式语体）	词义
about	approximately	大约
ask	inquire	询问
begin	commence	开始,着手
buy	purchase	买,购买
change	transform	转换,改变
cheap	inexpensive	便宜的
finish	complete	完成
get	obtain	获得,得到
give	accord	给予
have	possess	占有,拥有
method	technique	技术,方法
quick	rapid	迅速的
try	endeavor	努力,尽力
use	employ	使用
fire	flame	火焰

(Continued)

普通英语词（一般语体）	科技英语中的非技术词（正式语体）	词义
happy	excited	兴奋的
careful	caution	小心
heart	center	中心，中央
enough	sufficient	充分的，足够的
in the end	eventually	最后，终于

2. 一些短语动词往往由正式动词代替，表现出科技英语行文要求精练，表达上力避烦冗。例如：

短语动词	正式动词	词义
take in	absorb	吸收，吸引
push in	insert	插入
put up	erect	使直立，树立
put out	extinguish	熄灭
wear away	erode	侵蚀，腐蚀
take away	remove	移动，迁移
use up	consume	消耗，消费
carry out	perform	履行，执行
come across	encounter	遭遇，遇到

Lesson 5
Elements, Compounds and Mixtures
单质、化合物和混合物

Warming up

物质的分类：

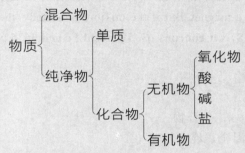

单质： 由同种元素组成的纯净物。

元素以单质形式存在时的状态称为元素的游离态。一般来说，单质的性质与其元素的性质（尤其是化学性质）密切相关。

注意： "单质"和"元素"在英语是同一个单词"Element"，不少英语学习者会把这两个概念混淆。有时为了区别，在强调"单质"时可用"free element"。

化合物： 由多种元素组成的纯净物。

单质是由一种元素组成的纯净物，而化合物是由两种或两种以上的元素组成的纯净物；从微观范围看，单质由同种原子构成，化合物由不同种原子构成；从性质上看，单质不能发生分解反应，化合物可以发生分解反应。

混合物： 混合物是由两种或多种物质混合而成的物质。

混合物没有固定的化学式，无固定组成和性质，组成混合物的各种成分之间没有发生化学反应，它们将保持着原来的性质。混合物可以用物理方法将所含物质加以分离。

单质和化合物都属于纯净物。判断物质是单质还是化合物，首先看物质是不是纯净物，只有属于纯净物才有可能属于单质或化合物。不能认为由同种元素组成的物质一定就是单质，也不能认为由不同种元素组成的物质一定是化合物。例如白磷和红磷，虽然都由磷元素组成，但它们不属于一种物质，混合后属于混合物，不属于纯净物。

Text

Most substances can be decomposed into two or more other substances. For example, water can be decomposed into hydrogen and oxygen. Table salt is easily decomposed into sodium and chlorine. However, an element can not be decomposed into simpler substances.

Compounds are composed of two or more elements. So they can be decomposed into simpler substances by chemical changes. A molecule is a small unit of a compound. If we divide a drop of water into smaller and smaller particle, we obtain a water molecule at last. A water molecule is composed of two hydrogen atoms and one oxygen atom. We cannot divide it if we don't destroy the molecule.

What are the characteristics of a mixture? If we mix the two elements sulfur and iron, do we have a compound? No, we have a mixture of the two elements. In fact, the iron and sulfur of the mixture can be separated by a magnet. But if the mixture is heated, the iron and sulfur combine to form iron (Ⅱ) sulfide (FeS). It contains 63.5 percent Fe and 36.5 percent S by weight. It is not attracted by a magnet.

New Words

compound ['kɒmpaʊnd] *n.* 化合物
mixture ['mɪkstʃə] *n.* 混合，混合物
decompose [ˌdi:kəm'pəʊz] *v.* 分解
oxygen ['ɒksɪdʒən] *n.* 氧，氧气
sodium ['səʊdɪəm] *n.* 钠
chlorine ['klɔ:ri:n] *n.* 氯，氯气
molecule ['mɒlɪkju:l] *n.* 分子
unit ['ju:nɪt] *n.* 个体，单位
divide [dɪ'vaɪd] *v.* 分，划分，分开
particle ['pɑ:tɪkl] *n.* 粒子，微粒
obtain [əb'teɪn] *vt.* 获得，得到
atom ['ætəm] *n.* 原子
destroy [dɪ'strɔɪ] *vt.* 破坏，毁坏，消灭
mix [mɪks] *v.* 使混合，混合
sulfur ['sʌlfə] *n.* 硫，硫黄
form [fɔ:m] *n.* 形状；*v.* 形成，构成
iron ['aɪən] *n.* 铁
magnet ['mægnət] *n.* 磁铁
attract [ə'trækt] *vt.* 吸引，有吸引力

Expressions and Technical Terms

be decomposed into 被分解为……
table salt 食盐

be composed of 由……组成
be separated 被分离
iron（Ⅱ）sulfide（FeS）硫化亚铁

Exercises

A. Translate the following into English.

1. 氢气　　2. 氧气　　3. 食盐　　4. 钠

5. 氯气　　6. 分子　　7. 原子　　8. 铁

9. 硫　　　10. 硫化亚铁

B. Decide whether the following statements are true（T）or false（F）. Write T for true and F for false in each blank.

（　）1. All substances can be decomposed into two or more other substances.
（　）2. An element can not be decomposed into simpler substances.
（　）3. Because compounds are composed of two or more elements, they can be decomposed into simpler substances by chemical changes.
（　）4. We can divide a water molecule if we don't destroy the molecule.
（　）5. If we mix the two elements sulfur and iron, we have a compound.

C. Translate the following sentences into Chinese.

1. A water molecule is composed of two hydrogen atoms and one oxygen atom.

2. Iron（Ⅱ）sulfide（FeS）contains 63.5 percent Fe and 36.5 percent S by weight.

Reading Material　阅读材料

科技英语文体的主要特点（二）

在科技英语文体中，句子的谓语动词常以被动语态的形式出现。如：
1. Air and water can be converted into nitric acid.
（空气和水能被转变成硝酸）
2. It is called chemical reactions.（它被称为化学反应）

虽然主动结构和被动结构意义相近，但被动结构使人一目了然。首先，被动句并不提及人，对于一个科学家或科技工作者来说，过多的提及人不但没有必要，而且会引起误会。其次主语是句子中非常重要的部分，把不提及人的这一部分放在句首，能引起读者的注意。在科技英语文体中谓语动词用被动语态形式可使句子简洁。科技英语中很多常用的被动语态结构在汉语中已有习惯的译法。如：

It is considered that... 人们认为……
It is supposed that... 据推测，假定……
It is noticed that... 人们注意到……
It has been shown that... 已经表明……
It is reported that... 据报道……
be known as... 通常为……，叫作……
be considered as... 被说成是……
be described as... 被描述为……
be defined as... 被定义为……

Lesson 6
Solids, Liquids and Gases
固体、液体和气体

Warming up

固态： 物质存在的一种热力学平衡状态。与液体和气体相比，固体有比较固定的体积和形状，质地比较坚硬。

组成固体的微粒间距离很小，作用力很大。粒子在各自的平衡位置附近作无规律的振动，固体能保持一定的体积和形状，流动性差，一般不存在自由移动粒子，它们的导电性通常由自由移动电子引起的。在受到不太大的外力作用时，固体的体积和形状改变很小。

固体分为晶体和非晶体，晶体具有固定的熔化温度，非晶体没有固定的熔化温度。

液态： 指物质的液体状态，液体可以流动、变形，可微压缩。

液态时，分子间主要起作用的力是范德华力。范德华力是由分子间的偶极矩相互吸引而形成的。所以不像化学键有固定的角度，范德华力只有大概的方向。这也是液体为什么会流动而固体不能的原因。

液体与气体不同，它有一定的体积。液体又与固体不同，它有流动性，因而没有固定的形状。与气体不同，液体的黏性较气体大，且随温度的升高而降低。

气态： 物质的一个状态。气体与液体一样是流体：它可以流动，可变形。与液体不同的是气体可以被压缩。假如没有限制（容器或力场）的话，气体可以扩散，其体积不受限制。气态物质的原子或分子相互之间可以自由运动。气态物质的原子或分子的动能比较高。

气体形态可通过改变其体积、温度和压强而改变。

理想气体： 为假想的气体。其特性为：气体分子间无作用力，气体分子本身不占有体积，气体分子与容器器壁间发生完全弹性碰撞。真实气体在愈低压、愈高温的状态，性质愈接近理想气体。最接近理想气体的气体为氦气。

Text

Water would be described as a colourless liquid. In very cold conditions it becomes a solid called ice, and when heated on a fire it becomes a vapour called steam. But water, people would say, is a liquid.

Water has been known to consist of molecules composed of two atoms of hydrogen and one atom of oxygen, which is described by the formula H_2O. But this is equally true of the solid called ice and the gas called steam. Chemically there is no difference between the gas, the liquid, and the solid, all of which are made up of molecules with the formula H_2O. And this is true of other chemical substances; most of them can exist as gases or as liquids or as solids. For instance, iron—a very hard solid, will melt into liquid when it is heated to a very high temperature while air—a gas, will become liquid when it is very cold.

Nothing very permanent occurs when a gas changes into a liquid or a solid. Ice, which has been made by freezing water, is known to be able to melted again by being warmed while steam can be condensed on a cold surface to become liquid water. In fact, it is only because water is such a familiar substance that different names are used for the solid, liquid and gas. For other substances, these different states have to be described directly. Thus for air, such names as liquid air, solid air and gaseous air are often mentioned about, but, since gaseous air is the normal thing, it is usually just described as air.

What, then, does it mean that water is called a liquid, air a gas and salt a solid? It means nothing more than that this is the usual condition of things on the earth. On one of the outer planets all three substances would be gases. Most substances are only familiar to be in one state, because the temperatures required to turn them into gases are very high, or the temperatures necessary to turn them into solids are so low. Water is an exception in this respect, which is another reason why its three states have been given three different names.

The fact that a liquid can be changed to a solid and back again to the original state, just by changing the temperature, shows that the very strong bonds between the atoms in the molecules have not been greatly changed. The difference between these three forms lies simply in the arrangement of the molecules or their position with respect to each other.

❖ New Words

describe [dɪˈskraɪb] *vt.* 把……说成，形容，描写
colo(u)rless [ˈkʌlələs] *adj.* 无色的，苍白的
compose [kəmˈpəʊz] *vt.* 组成，构成
atom [ˈætəm] *n.* 原子
formula [ˈfɔːmjələ] *n.* 化学式，公式，配方
exist [ɪɡˈzɪst] *vi.* 存在，生存
instance [ˈɪnstəns] *n.* 例子，实例
melt [melt] *v.* 熔化，溶解，使熔化
permanent [ˈpɜːmənənt] *adj.* 永久的，持久的
occur [əˈkɜː(r)] *vi.* 出现，发生，存在
freezing [ˈfriːzɪŋ] *adj.* 冻结的，凝固的
gaseous [ˈɡæsɪəs] *adj.* 气体的，气态的
mention [ˈmenʃn] *vt.* 提到，说起；*n.* 提到
outer [ˈaʊtə(r)] *adj.* 外部的，外侧的

exception [ɪkˈsepʃn] *n.* 例外，除外
respect [rɪˈspekt] *n.* 方面
original [əˈrɪdʒənl] *adj.* 原先的，最初的
bond [bɒnd] *n.* 键，联结，结合物，黏结剂，债券
arrangement [əˈreɪndʒmənt] *n.* 排列，安排

Expressions and Technical Terms

describe as... 把……说成
be composed of... 由……组成
with respect to... 关于……
in respect of... 就……而言

Exercises

A. Translate the following into English.
1. 固态空气
2. 原子间极强的键
3. 两个氢原子
4. 分子相互之间的位置
5. 以液体的形态存在
6. 高温

B. Decide whether the following statements are true (T) or false (F). Write T for true and F for false in each blank.

(   ) 1. The three states of water have three names because the formula H_2O will be changed with their different forms.

(   ) 2. The solid air is colourless.

(   ) 3. A bar of copper will change into a liquid form if it is heated to a high temperature.

(   ) 4. A molecule of water is made up of two atoms of hydrogen and one atom of oxygen.

(   ) 5. The word "exception" in Para. 4 means that the three states of water are so familiar to us that they have different names.

(   ) 6. The reason why we seldom speak of gaseous air is that there is no such thing in the world.

C. Cloze.
Matter may be in three different states—solid, liquid and gas. Some kind of matter may be easily _____ from one state into _____ . Water is a _____ , but when its temperature gets lower than _____ , it will freeze into ice and become a _____ . And when water is _____ to 100℃, it will form steam and become a _____ . So ice is the solid _____ of water and steam is _____ in the form of gas. Gas and solid can also be changed into _____ .

Lesson 7
Nomenclature of Inorganic Compounds
无机物命名法

Warming up

一、元素和单质的命名

"元素"和"单质"的英文意思都是"element",有时为了区别,在强调"单质"时可用"free element"。因此,单质的英文名称与元素的英文名称是一样的。

二、阳离子

1. 单价阳离子:直呼其名,即读其元素名称。
2. 多价阳离子:直呼其名后再加罗马数字表示价态。

$$Mn^{4+}: Manganese(Ⅳ); Mn^{2+}: Manganese(Ⅱ)$$

对于有变价的金属元素,除了可用前缀来表示外,更多采用罗马数字来表示金属的氧化态,或用后缀-ous 表示低价,-ic 表示高价。

$$FeO: iron(Ⅱ) \text{ oxide or ferrous oxide}$$
$$Fe_2O_3: iron(Ⅲ) \text{ oxide or ferric oxide}$$

三、阴离子

1. 二元化合物

(1)常见的二元化合物有卤化物,氧化物,硫化物,氮化物,磷化物,碳化物,金属氢化物等,命名时需要使用后缀-ide,OH^- 的名称也是用后缀-ide: hydroxide。

如:对于金属氧化物是在阳离子名称后加 oxide;

对于金属氢氧化物则是在阳离子名称后加 hydroxide;

对于非金属氧化物,命名规则是在非金属元素名称前和 oxide 前加表示数目的前缀。

CO_2 Carbon dioxide; N_2O_3 Dinitrogen trioxide

(2)非金属氢化物不用此后缀,而是将其看成其他二元化合物。只有水和氨这两种非金属氢化物经常使用俗称 H_2O: water; NH_3: ammonia。除上所述外,其他的非金属氢化物都用系统名称,命名规则根据化学式的写法不同而有所不同。

① 对于卤族和氧族氢化物,H 在化学式中写在前面,因此将其看成与另一元素的二元化合物。 HF: hydrogen fluoride; HCl: hydrogen chloride。

② 对于其他族的非金属氢化物,H 在化学式中写在后面,可加后缀-ane,氮族还可加-ine。 PH_3: phosphine 或 phosphane; AsH_3: arsine 或 arsane。

（3）非最低价的二元化合物还要加前缀，如：peroxide 过氧化物；superoxide 超氧化物。

上述提到的表示数目的前缀为希腊语前缀，几乎每个英文数字前缀都同时对应有希腊语和拉丁语两个前缀，如下所示：

数字	希腊语	拉丁语
一	mono-['mɒnəʊ]	uni-['ju:nɪ]
二	di-[daɪ]	bi-[baɪ]
三	tri-[traɪ]	ter-[tɜ:]
四	tetra-['tetrə]	quadri-['kwɒdrɪ]
五	penta-[pentə]	quint-[kwɪnt]
六	hexa-['heksə]	sex-[seks]
七	hepta-['heptə]	sept-[sept]
八	octa-[ɒk'tə]	
九	ennea-['enɪə]	
十	deca-['dekə]	
十一	hendeca-['hen dekə]	
十二	dodeca-['dəʊ'nɪkə]	

英语里表示数目的前缀，一般来自拉丁语和希腊语两种，在不同场合，各有其用处。通常情况下，带有拉丁语前缀的词，大多是用来指物质世界或现实生活的东西；而表示思想感情的用语，或数学、艺术与哲学的词汇，则多为希腊语前缀。

2. 无氧酸

命名规则：hydro-词根-ic acid，如：HCl　hydrochloric acid；H_2S　hydrosulfuric acid。

3. 含氧酸与含氧酸根阴离子（以正酸和亚酸为例）

（1）正酸与正酸根

命名规则：正酸-ic；正酸根-ate。如：$HClO_3$　chloric acid；ClO_3^-　chlorate。

（2）亚酸与亚酸根

命名规则：亚酸-ous；亚酸根-ite。如：H_2SO_3　sulfurous acid；SO_3^{2-}　sulfite。

四、盐（以正盐和酸式盐为例）

1. 正盐

根据化学式从左往右分别读出阳离子和阴离子的名称。

normal salt = cation + anion

如：Na_2CO_3　sodium carbonate。

2. 酸式盐

同正盐的读法，酸根中的 H 读作 hydrogen，氢原子的个数用前缀表示。

acidic salt = cation + hydrogen + anion

如：NaH_2PO_4　sodium dihydrogen phosphate。

Text

The systematic name of inorganic compounds considers the compound to be composed of two parts, one positive and one negative. The positive part is named and written first. The negative part, generally nonmetallic, is named second. Names of the elements are modified with suffixes and prefixes to identify the different types of compounds.

Examples:

Symbol(符号)	Element(元素)	Stem(词根)	Binary name(某化物)
F	fluorine	fluor-	fluoride(氟化物)
Cl	chlorine	chlor-	chloride(氯化物)
Br	bromine	brom-	bromide(溴化物)
I	iodine	iod-	iodide(碘化物)
O	oxygen	ox-	oxide(氧化物)
N	nitrogen	nitr-	nitride(氮化物)
S	sulfur	sulf- or sulfur-	sulfide(硫化物)
P	phosphorus	phosph-	phosphide(磷化物)
C	carbon	carb-	carbide(碳化物)
H	hydrogen	hydr-	hydride(氢化物)

Binary compounds

Binary compounds contain only two different elements. Their name consists of two parts, the name of electropositive element and the name of electronegative element which is modified to end in ide.

(1) Binary compounds containing a metal and nonmetals

Examples:

Formula (化学式)	Name (名称)
Na_2O	sodium oxide
CaC_2	calcium carbide
$AlCl_3$	aluminum chloride
NH_4F	ammonium fluoride
PbS	lead sulfide

(2) Binary compounds containing two nonmetals

Using Latin prefix to indicate the number of atoms in the molecule.

mono=1　　di=2　　tri=3　　tetra=4　　penta=5
hexa=6　　hepta=7　　octa=8　　nona=9　　deca=10

Examples:

Formula（化学式）	Name（名称）
CO	carbon monoxide
CO_2	carbon dioxide
PCl_3	phosphorus trichloride
PCl_5	phosphorus pentachloride
N_2O_4	dinitrogen tetroxide
NO	nitrogen oxide
N_2O_3	dinitrogen trioxide
SO_2	sulfur dioxide

Acids containing no oxygen atom in the molecule

Examples：

Formula（化学式）	非含氧酸（hydro＋词根＋ic acid）
HF	hydrofluoric acid
HCl	hydrochloric acid
HBr	hydrobromic acid
HI	hydroiodic acid
H_2S	hydrosulfuric acid

Name of oxy-acids and oxy-compounds

Examples：

Formula（化学式）	含氧酸（词根＋ic acid）	对应盐（词根＋ate）
H_2SO_4	sulfuric acid	sulfate
HNO_3	nitric acid	nitrate
H_2CO_3	carbonic acid	carbonate
$HBrO_3$	bromic acid	bromate
亚含氧酸（词根＋ous acid）		对应盐（词根＋ite）
H_2SO_3	sulfurous acid	sulfite
HNO_2	nitrous acid	nitrite

Name of salts

Examples：

Formula（化学式）	Name of salt（盐名称）
Na_2SO_3	sodium sulfite
Na_2SO_4	sodium sulfate
KNO_2	potassium nitrite
KNO_3	potassium nitrate
$CaCO_3$	calcium carbonate

(Continued)

Formula(化学式)	Name of salt(盐名称)
MgBrO₃	magnesium bromate

Name of bases

Examples:

Formula(化学式)	Name of base(hydroxide)碱名称(氢氧化物)
KOH	potassium hydroxide
NH₃·H₂O	ammonium hydroxide
Ca(OH)₂	calcium hydroxide

New Words

nomenclature [nəˈmeŋklətʃə(r)] n. 命名法
positive [ˈpɒzətɪv] adj. 正的，阳性的
negative [ˈnegətɪv] adj. 负的，阴性的
nonmetallic [ˌnɒnmɪˈtælɪk] adj. 非金属的
modify [ˈmɒdɪfaɪ] vt. 更改，修改
suffix [ˈsʌfɪks] n. 后缀
prefix [ˈpriːfɪks] n. 前缀
fluorine [ˈflɔːriːn] n. 氟
binary [ˈbaɪnərɪ] adj. 二元的
bromine [ˈbrəʊmiːn] n. 溴
iodine [ˈaɪədiːn] n. 碘
phosphorus [ˈfɒsfərəs] n. 磷
calcium [ˈkælsɪəm] n. 钙
aluminum [əˈluːmɪnəm] n. 铝
ammonium [əˈməʊnɪəm] n. 铵
lead [liːd] n. 铅
acid [ˈæsɪd] n. 酸
potassium [pəˈtæsjəm] n. 钾
magnesium [mægˈniːzɪəm] n. 镁
base [beɪs] n. 碱
hydroxide [haɪˈdrɒksaɪd] n. 氢氧化物

Expressions and Technical Terms

phosphorus trichloride 三氯化磷
phosphorus pentachloride 五氯化磷
dinitrogen tetroxide 四氧化二氮
nitrogen oxide 一氧化氮

dinitrogen trioxide 三氧化二氮
sulfur dioxide 二氧化硫
hydrofluoric acid 氢氟酸
hydrochloric acid 盐酸
hydrobromic acid 氢溴酸
hydroiodic acid 氢碘酸
hydrosulfuric acid 氢硫酸
sulfuric acid 硫酸
nitric acid 硝酸
carbonic acid 碳酸
bromic acid 溴酸
sulfurous acid 亚硫酸
nitrous acid 亚硝酸
sodium sulfite 亚硫酸钠
potassium nitrate 硝酸钾
potassium hydroxide 氢氧化钾
calcium hydroxide 氢氧化钙

Exercises

A. Fill in the blanks.

Symbol(符号)	Element(元素)	Stem(词根)	Binary name(某化物)
Cl	chlorine		chloride(氯化物)
O	oxygen	ox-	（氧化物）
N		nitr-	nitride(氮化物)
S	sulfur		（硫化物）
C			（碳化物）
H	hydrogen		（氢化物）

Formula(化学式)	Name(名称)
Na_2O	
	calcium carbide
	ammonium fluoride

Formula(化学式)	Name(名称)
	carbon monoxide
	carbon dioxide
NO	
	dinitrogen trioxide
SO_2	

Formula(化学式)	非含氧酸(hydro＋词根＋ic acid)
	hydrofluoric acid
HCl	
	hydrobromic acid
	hydrosulfuric acid

Formula(化学式)	含氧酸(词根＋ic acid)	对应盐(词根＋ate)
H_2SO_4		sulfate
HNO_3		nitrate
H_2CO_3		carbonate

Formula(化学式)	Name of salt(盐名称)
Na_2SO_4	
	potassium nitrate
$CaCO_3$	

Formula(化学式)	Name of base(hydroxide)碱名称(氢氧化物)
	potassium hydroxide
	ammonium hydroxide
$Ca(OH)_2$	

B. Translate the following into Chinese.

1. sulfuric acid
2. nitrogen oxide
3. potassium carbonate
4. calcium hydroxide
5. hydrochloric acid
6. dinitrogen trioxide
7. sodium nitrite
8. sodium nitrate
9. ammonium hydroxide
10. hydrosulfuric acid
11. carbon monoxide
12. calcium hydroxide

Reading Material 阅读材料

科技英语构词法

1. 转化法

由一种词类转化成另一种词类，叫转化法。

例如：

water（n. 水）→water（v. 浇水）

charge（n. 电荷）→charge（v. 充电）

2. 派生法

通过加前、后缀构成一个新词，叫派生法。派生法是化工类科技英语中最常用的构词法。例如："烷烃"就是用前缀表示分子中碳原子数再加上"ane"作词尾构成的。如下表：

常见有机物构词规律

烷烃（后缀 ane）	化学名称	烷烃（后缀 ane）	化学名称
methane	甲烷	ethane	乙烷
propane	丙烷	butane	丁烷
pentane	戊烷	hexane	己烷
heptane	庚烷	octane	辛烷
tridecane	十三烷	tetradecane	十四烷
pentadecane	十五烷	hexadecane	十六烷
烷基取代基（后缀 yl）	化学名称	烷基取代基（后缀 yl）	化学名称
methyl	甲基	ethyl	乙基
propyl	丙基	butyl	丁基
pentyl	戊基	hexyl	己基
烯烃（后缀 ene）	化学名称	炔烃（后缀 yne）	化学名称
ethene	乙烯	ethyne	乙炔
propene	丙烯	propyne	丙炔
butene	丁烯	butyne	丁炔
pentene	戊烯	pentyne	戊炔
醇（alcohol）	化学名称	醚（ether）	化学名称
methyl alcohol	甲醇	methyl ether	甲醚
ethyl alcohol	乙醇	ethyl ether	乙醚
propyl alcohol	丙醇	propyl ether	丙醚

有机物的系统命名中以表示碳数多少的拉丁或希腊数字前缀，结合表示有机物官能团特征后缀构成，其中主要后缀及其名称有：-ane 烷、-ene 烯、-yne 或-ine 炔、-yl 基、-anol/alcohol 醇、phenol 酚、carboxylic acid 羧酸或-anoic acid 酸、acid anhydride 酸酐、-ose 糖、-ase 酶、-oside 糖苷、-anal/aldehyde 醛、-anone/ketone 酮、ester 酯、ether 醚、lactone 内酯、-amine 胺、-amide 酰胺、-azide 叠氮、hydrazine 肼、hydrazone 腙；phenyl 苯基、carbonyl 羰基、carboxyl 羧基、benzyl 苄基、aryl 芳基等。

3. 合成法

由两个或更多的词合成一个词，叫合成法。有时需要加连字符。

如：名词＋名词　　carbon steel　　　　碳钢
　　　　　　　　　rust-resistance　　　防锈
　　形容词＋名词　atomic weight　　　　原子量
　　　　　　　　　periodic table　　　　周期表

4. 压缩法

（1）只取词头字母

CET　　　　　　　　　College English Test　　大学英语考试

CAD	Computer Aided Design	计算机辅助设计

（2）将单词删去一些字母

Lab	Laboratory	实验室
Corp	Corporation	股份公司
Exam	Examination	考试
Kilo	Kilogram	千克，公斤

Lesson 8
Acids and Bases 酸和碱

Warming up

酸：化学上是指在水溶液中电离时产生的阳离子都是氢离子的化合物。可分为无机酸、有机酸。酸碱质子理论认为，能释放出质子的物质总称为酸。

碱：在酸碱电离理论中，碱指在水溶液中电离出的阴离子全部都是 OH^- 的物质；在酸碱质子理论中碱指能够接受质子的物质。

酸的化学性质：
1. 能和碱或碱性氧化物反应生成盐和水；
2. 能与某些金属反应生成盐和氢气；
3. 水溶液有酸味并能使指示剂变色，如紫色石蕊变红。

碱的化学性质：
1. 碱溶液能与酸碱指示剂作用；
2. 碱能与非金属单质发生反应；
3. 碱能与酸发生反应，生成盐和水；
4. 碱溶液能与酸性氧化物反应，生成盐和水；
5. 碱溶液（相对强碱）能与盐反应，生成新碱（相对弱碱）和新盐。

Text

You run into dozens of acids and bases every day. Orange juice contains acid. But the soap you use is a base. Believe it or not, there are acids and bases in your own body. Your stomach makes acid that helps you digest food. Acids are key ingredients in your muscles and skin. Your tears contain a base. So does your blood.

What makes an acid different from a base? The answer has to do with ions. When two chemical substances interact, atoms of one substance may lose negative charges to the other. The atom that loses a negative charge becomes a positive ion. The atom that gains a negative charge becomes a negative ion.

An acid is a substance that dissolves in water in a way that releases positive hydrogen ions (H^+) into the solution. A base is a substance that dissolves in water in a way that releases hydroxide ions (OH^-) in the solution. But ions are too small to see, even with a microscope. Are there other ways to tell if something is an acid or a base? Yes indeed!

Acids corrode metal. In many parts of the world, acid rain falls from the sky. When coal and other fossil fuels are burned, chemicals are released into the air. Inside the clouds, some of these chemicals combine with water to form acids. They fall to Earth as acid rain (or acid snow). Acids corrode metal structures. It also dissolves some kinds of stone and kills living things such as trees and fish. Acids taste sour.

Bases taste bitter. Bases feel slippery when you touch them. They are good at dissolving certain substances, like grease. That's one reason why bases make good cleaning agents. But bases also break down proteins. Some bases can actually dissolve your skin. Acids and bases are everywhere, but some of them need to be used with care.

New Words

juice [dʒuːs] *n.* （水果）汁，液
stomach [ˈstʌmək] *n.* 胃，胃部
digest [daɪˈdʒest] *v.* 消化
key [kiː] *n.* 关键
ingredient [ɪnˈɡriːdiənt] *n.* 成分
interact [ˌɪntərˈækt] *vi.* 互相作用，互相影响
corrode [kəˈrəʊd] *v.* 使腐蚀，侵蚀
fossil [ˈfɒsl] *adj.* 化石的
sour [ˈsaʊə(r)] *adj.* 酸的，酸味的
bitter [ˈbɪtə(r)] *adj.* 苦的
grease [ɡriːs] *n.* 油脂
protein [ˈprəʊtiːn] *n.* 蛋白质

Expressions and Technical Terms

run into 遇到
dozens of 几十，许多
believe it or not 无论相信与否
make… different from… 使……和……不同
do with 处理
acid rain 酸雨
acid snow 酸雪
fossil fuel 化石燃料
break down 分解
with care 小心地

Exercises

A. Translate the following into English.

1. 酸　　　　2. 碱　　　　　3. 电荷　　　　4. 阳离子

5. 阴离子　　6. 显微镜　　　7. 腐蚀　　　　8. 酸雨

B. Choose the best answer to each question.

1. Your tears and your blood contains （　　）
（A）an acid.　　　　　　　（B）a base.
2. The atom that loses a negative charge becomes （　　）
（A）a positive ion.　　　　（B）a negative ion.
3. When an acid dissolves in water, it can releases （　　）
（A）hydroxide ions.　　　　（B）positive hydrogen ions.
4. Which of the following is true? （　　）
（A）Bases taste sour.
（B）Bases make good cleaning agents.
（C）Acid rain can not corrode metal.

C. Cloze.

What makes an acid different from a base? The answer has to do with _____ . When two chemical substances interact, atoms of one substance may lose _____ charges to the other. The substance that loses a negative charge becomes a _____ ion. The substance that gains a negative charge becomes a _____ ion.

Reading Material　阅读材料

Description of Acids and Bases

Acids and bases can be dangerous. How can you figure out if the liquids are acids or bases without touching or smelling them? The easiest way is to use an indicator. An indicator is something that changes color when it is exposed to an acid or a base. One of the most common acid/base indicators is litmus paper. This is paper that has been soaked in a mixture of chemicals called litmus. Litmus may be the world's oldest acid/base indicator. It has been used for hundreds of years. Litmus paper has two forms—blue and red. If you dip a strip of blue litmus paper into a solution and it turns red, you know you have got an acid. If you dip a strip of red litmus paper into a solution and it turns blue, you have got a base. It is that simple.

But there is something more you need know. How strong are the acids or the bases? If you spilled apple juice on your skin, you would not worry. But if you spilled sulfuric acid on your skin, you would get a terrible burn. Both these liquids are acids. But there is a big difference between them. Sulfuric acid is much more powerful or stronger acid than apple juice.

Whether an acid is strong or weak depends on how many hydrogen ions it releases when it dissolves in water. The more hydrogen ions there are in the solution, the stronger the acid is. Similarly, the strength of a base depends on how many hydroxide ions there are in the solution. The more hydroxide ions there are, the stronger the base is. Scientists measure the strength of an acid or a base. This measurement is called the pH. A set of pH number is from 0 to 14. Pure water is in the middle, with a pH of 7. That means pure water is neutral, which is neither an acid nor a base.

What happens when you mix equal amounts of an acid and a base of equal strength? They cancel each other to make a neutral solution. Scientist calls this reaction "neutralizing."

New Words

indicator [ˈɪndɪkeɪtə(r)] n. 指示剂
expose [ɪkˈspəʊz] v. 使暴露，揭露
litmus [ˈlɪtməs] n. 石蕊
soak [səʊk] v. 浸，泡，浸透
strip [strɪp] n. 条，带
spill [spɪl] v. 溢出，溅出
neutral [ˈnjuːtrəl] adj. 中性的

Expressions and Technical Terms

figure out 解决，弄明白，计算出
litmus paper 石蕊试纸
a strip of 一长条
sulfuric acid 硫酸
pH test paper pH 试纸
equal amount of... 等量的……
each other 互相

Exercises

A. Translate the following into English.

1. 计算出　　　　2. 指示剂　　　　3. 石蕊试纸　　　　4. pH 试纸

5. 硫酸　　　　　6. 测量　　　　　7. 混合　　　　　　8. 中性的

B. Choose the best answer to each question.

1. When a strip of blue litmus paper turns red in a solution, that means the solution is (　　)
(A) an acid solution.
(B) a base solution.

(C) a neutral solution.

2. Sulfuric acid is (　　)

(A) much more powerful than apple juice.

(B) weaker than apple juice.

3. When there are more hydrogen ions in the solution, it means (　　)

(A) the weaker the acid is.

(B) the stronger the acid is.

(C) the solution is a neutral solution.

Lesson 9
Temperature 温度

Warming up

温度： 表示物体冷热程度的物理量，微观上来讲是物体分子热运动的剧烈程度。从分子运动论观点看，温度是物体分子运动平均动能的标志。温度是大量分子热运动的集体表现，含有统计意义。

温度只能通过物体随温度变化的某些特性来间接测量，而用来量度物体温度数值的标尺叫温标。它规定了温度的读数起点（零点）和测量温度的基本单位。

温度的国际单位： 为热力学温标（K），国际上用得较多的其他温标有华氏温标（℉）、摄氏温标（℃）和国际实用温标。

气温： 大气层中气体的温度是气温，是气象学常用名词。它直接受日射影响。中国以摄氏温标（℃）表示。气象部门所说的地面气温，就是指高地面约 1.5m 处百叶箱中的温度。

摄氏温标： 1740 年瑞典天文学家摄尔修斯（Celsius）提出在标准大气压下，把冰水混合物的温度规定为 0 度，水的沸腾温度规定为 99.974 度。根据水这两个固定温度点来对玻璃水银温度计进行分度。两点间作 100 等分，每一份称为 1 摄氏度，记作 1℃。摄氏温度已被纳入国际单位制。物理学中摄氏温度表示为 t，绝对温度（单位：开尔文）表示为 T，摄氏温度的定义是 $t = T - 273.15$。

温标间关系：

摄氏温度和华氏温度的关系：$T(℉) = 1.8t(℃) + 32$

摄氏温度和热力学温度的关系：$T(K) = t(℃) + 273.15$

Text

When anything is hot, it is said to have a high temperature, and when it cold, it is said to have a low temperature. Temperature is measured in degrees with an instrument known as thermometer. The figures are determined by the temperature of melting ice and that of boiling water. These are known as the melting point of ice and the boiling point of water.

There are two important systems: degrees Kelvin (K) and degrees centigrade (℃). Ice is said to melt at 273 degrees Kelvin (273K) or at 0 degrees centigrade (0℃). Water is said to boil at 373 degrees Kelvin (373K) or at 100 degrees centigrade (100℃). In science, the centigrade system of measuring temperature is used.

A mercury thermometer consists of a glass tube having a bulb at one end. The bulb and lower part of the tube contain mercury. In order to determine the figures on a thermometer, the bulb is placed in melting ice and a mark is made at the level of the mercury, calling this point 0. Then the bulb is placed in steam and the mercury rises in the tube, because it expands on heating. Another mark is made at the new level of the mercury, calling this point 100. The space between the two marks is divided into 100 equal divisions for a centigrade thermometer.

New Words

temperature ['temprətʃə(r)] *n.* 温度
measure ['meʒə(r)] *vt.* 测量，测定
thermometer [θə'mɒmɪtə(r)] *n.* 温度计
Kelvin ['kelvɪn] *adj.* 开尔文温度计的
centigrade ['sentɪgreɪd] *adj.* 摄氏温度计的
mercury ['mɜːkjərɪ] *n.* 水银（柱），汞
bulb [bʌlb] *n.* 球状物，玻璃泡（球管）
expand [ɪk'spænd] *v.* 膨胀
division [dɪ'vɪʒn] *n.* 分开，划分

Expressions and Technical Terms

to be known as… 叫作，称为……
melting point 熔点
boiling point 沸点
degrees Kelvin 热力学温标，开尔文温度
degrees centigrade 摄氏温标，摄氏度

Exercises

A. Translate the following into English.

1. 温度计　　　　2. 熔点　　　　3. 沸点　　　　4. 开尔文温度

5. 摄氏度　　　　6. 水银　　　　7. 一百等分　　　8. 温度

B. Decide whether the following statements are true (T) or false (F). Write T for true and F for false in each blank.

(　　) 1. When anything is hot, it is said to have a low temperature, and when it cold, it is said to have a high temperature.

(　　) 2. Temperature is measured in degrees with an instrument known as thermometer.

(　　) 3. A mercury thermometer consists of a glass tube having a bulb at one end.

(　　) 4. There are two important systems: degrees Kelvin (K) and degrees centigrade (℃).

(　　) 5. Ice is said to melt at 273 degrees Kelvin (273K) or at 0 degrees centigrade (0℃). Water is said to boil at 373 degrees Kelvin (373K) or at 100 degrees centigrade (100℃).

C. Cloze.

1. The _____ are determined by the temperature of melting ice and that of boiling water. These are known as the _____ _____ of ice and the _____ _____ of water.

2. The bulb of mercury thermometer is placed in steam and the _____ rises in the tube, because it expands on heating. Another mark is made at the new level of the mercury, calling this point 100. The space between the two marks is divided into 100 _____ divisions for a centigrade thermometer.

Lesson 10
Thermometer 温度计

Warming up

温度计: 是可以准确地判断和测量温度的工具,分为指针温度计和数字温度计等。

其设计的依据有:利用固体、液体、气体受温度的影响而热胀冷缩的现象;在定容条件下,气体(或蒸气)的压强因温度不同而变换;热电效应的作用;电阻随温度的变换而变换;热辐射的影响等。 常用温度计介绍如下。

(1) 指针式温度计: 是形如仪表盘的温度计,也称寒暑表,用来测室温,是用金属的热胀冷缩原理制成的。 它是以双金属片作为感温元件,用来控制指针。 双金属片通常是用铜片和铁片铆在一起,且铜片在左、铁片在右。 由于铜的热胀冷缩效果要比铁明显得多,因此当温度升高时,铜片牵拉铁片向右弯曲,指针在双金属片的带动下就向右偏转(指向高温);反之,温度变低,指针在双金属片的带动下就向左偏转(指向低温)。

(2) 数字体温计: 是利用温度传感器将(温度)转换成数字信号,然后通过显示器(如液晶、数码管、LED 矩阵等)显示出数字形式的温度,能快速准确地测量人体温度的最高值。 与传统的水银体温计相比,具有读数方便、测量时间短、测量精度高、能记忆并有提示音等优点。 尤其是数字体温计不含水银,对人体及周围环境无害,特别适合于医院、家庭使用。

(3) 玻璃管温度计: 玻璃管温度计是利用热胀冷缩的原理来实现温度的测量的。 由于测温介质的膨胀系数与沸点及凝固点的不同,我们常见的玻璃管温度计主要有:煤油温度计、水银温度计、红钢笔水温度计。 它的优点是结构简单,使用方便,测量精度相对较高,价格低廉。 缺点是测量上下限和精度受玻璃质量与测温介质的性质限制,且玻璃易碎。

2020 年 10 月 16 日,国家药监局在其网站发布《国家药监局综合司关于履行〈关于汞的水俣公约〉有关事项的通知》。《通知》明确要求,自 2026 年 1 月 1 日起,我国将全面禁止生产含汞体温计和含汞血压计产品。

水银温度计洒落出来的汞必须立即用滴管、毛刷收集起来,并用水覆盖(最好用甘油),然后在污染处撒上硫黄粉,无液体后(一般约一周时间)方可清扫。

Text

The ordinary form of mercury thermometer is used for temperatures ranging from $-40°F$ to

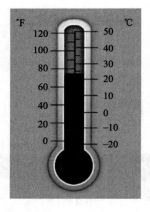

500°F. For measuring temperatures below －40°F, thermometers filled with alcohol are used. These are, however, not satisfactory for use at high temperatures. When a mercury thermometer is used for temperatures above 500°F, the space above the mercury is filled with some inert gases, usually nitrogen or carbon dioxide. As the mercury rises, the gas pressure is increased. So it is possible to use these thermometers for temperatures exceeding 1000°F. This is the limit, however, as the melting point of glass is comparatively low. For temperatures exceeding 800°F, some form of pyrometer is generally used. The simplest of these is the mechanical pyrometer.

New Words

mercury [ˈmɜːkjərɪ] n. 水银，汞
temperature [ˈtemprətʃə(r)] n. 温度，体温，气温
thermometer [θəˈmɒmɪtə(r)] n. 温度计，体温计
Fahrenheit [ˈfærənhaɪt] adj. 华氏温度计的，华氏的
measure [ˈmeʒə(r)] vt. 测量
alcohol [ˈælkəhɒl] n. 酒精，酒
space [speɪs] n. 空间
inert [ɪˈnɜːt] adj. 惰性的
exceed [ɪkˈsiːd] vt. 超越，胜过
comparatively [kəmˈpærətɪvlɪ] adv. 比较地，相当地
pyrometer [paɪˈrɒmɪtə(r)] n. 高温计

Expressions and Technical Terms

range from... to... 范围从……到……
fill... with... 用……充满……
inert gas 惰性气体
gas pressure 气体压力
melting point 熔点

Exercises

A. Translate the following into English.

1. 温度   2. 温度计   3. 水银   4. 酒精

5. 惰性气体   6. 气体压力   7. 熔点   8. 高温计

B. Choose the best answer to each question.

1. What topics is described in this passage? （   ）

（A） mercury thermometers.

（B） measuring temperatures.
（C） temperature range.

2. When we measure temperatures of about 300°F, which thermometer shall we choose? (   )

（A） pyrometer.
（B） mercury thermometer.
（C） inert gas thermometer.
（D） alcohol thermometer.

C. Translate the following sentences into Chinese.

1. For measuring temperatures below －40°F, thermometers filled with alcohol are used.

2. When a mercury thermometer is used for temperatures above 500°F, the space above the mercury is filled with some inert gas, usually nitrogen or carbon dioxide.

Reading Material 阅读材料

化学分析中常用仪器的名称（Instruments commonly used in chemical analysis）

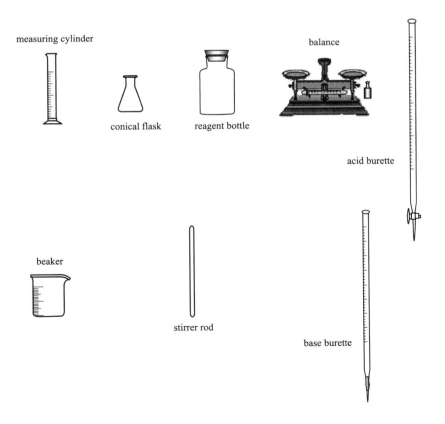

Lesson 11
Diffusion 扩散

 Warming up

扩散：物质分子从高浓度区域向低浓度区域转移，直到均匀分布的现象。

扩散现象是一个基于分子热运动（分子的无规则运动，称为热运动）的输运现象，是分子通过布朗运动从高浓度区域（或高化学势）向低浓度区域（或低化学势）运输的过程。

布朗运动：指悬浮在液体或气体中的微粒所做的永不停息的无规则运动。因其由英国植物学家布朗发现而得名。

扩散现象说明：一切物体的分子都在不停地做无规则运动。分子之间有间隙。

Text

The molecules of a substance are in continual motion. Imagine a crowd of people leaving a football match. They are all trying to get away as quickly as possible, and this is exactly how the molecules of a gas behave. They try to spread out, or in other words, to diffuse.

Diffusion can be described by opening a bottle of perfume in a room. The molecules of the perfume are in the bottle. Immediately the bottle is opened, they start to diffuse into the air, and the perfume can be smelt at the other end of the room. The longer the bottle is left open, the stronger the smell becomes.

Finally, when there is an equal concentration of molecules inside and outside the bottle, diffusion ceases. According to the principles of diffusion, the process continues until the concentration of perfume molecules is equal in all parts of the room.

New Words

motion [ˈməʊʃn] *n*. 运动，动作
crowd [kraʊd] *n*. 人群，群众
spread [spred] *v*. 伸展，展开
diffuse [dɪˈfjuːz] *v*. 散播，传播，散开
diffusion [dɪˈfjuːʒn] *n*. 扩散，传播
cease [siːs] *v*. 停止
principle [ˈprɪnsəpl] *v*. 原则，原理

Expressions and Technical Terms

a crowd of... 一群……
get away 离开
spread out 扩散开
in other words 换言之
according to the principles of... 根据……定律

Exercises

A. Translate the following into English.

1. 扩散   2. 分子   3. 原子   4. 离子

5. 扩散   6. 闻   7. 浓度   8. 根据……定律

B. Choose the best answer to each question.

1. The molecules of a substance (   )
(A) do not move.
(B) are in continual motion.
(C) move as slowly as possible.

2. The diffusion of the perfume molecules ceases, when (   )
(A) the concentration in the bottle is lower than that in the room.
(B) the concentration of the perfume molecules is equal in all parts of the room.

C. Decide whether the following statements are true (T) or false (F). Write T for true and F for false in each blank.

(   ) 1. The molecules of a substance are in continual motion.

(   ) 2. When there is an equal concentration of molecules inside and outside the bottle, diffusion begins.

(   ) 3. According to the principles of diffusion, the process continues until the concentration of perfume molecules is equal in all parts of the room.

Lesson 12
Catalysis 催化作用

Warming up

催化作用： 由于催化剂的介入而加速或减缓化学反应速率的现象。

在催化反应中，催化剂与反应物发生作用，改变了反应途径，从而降低了反应的活化能，这是催化剂得以提高反应速率的原因。

如：化学反应 A + B→AB，所需活化能为 E，加入催化剂 C 后，反应分两步进行，所需活化能分别为 F、G，其中 F、G 均小于 E。

催化作用可分为以下几种类型：①均相催化，②多相催化，③生物催化，④自动催化。 其他还有电催化、光助催化、光电催化……

催化剂： 在化学反应里能改变反应物化学反应速率（提高或降低）而不改变化学平衡，且本身的质量和化学性质在化学反应前后都没有发生改变的物质。

催化剂种类繁多，按状态可分为液体催化剂和固体催化剂；按反应体系的相态分为均相催化剂和多相催化剂。

Text

A catalyst is very important in chemical reactions. It can increase the rate of a reaction, but it is not used up itself in the reaction. For example, in the hydrolysis of an ester, hydroxide ion is not required in the reaction, but its addition increases the reaction rate.

$$CH_3COOC_2H_5 + H_2O \xrightleftharpoons{OH^-} C_2H_5OH + CH_3COOH$$

We can call the hydroxide ion a catalyst.

Now how do catalysts increase reaction rate? We can use activation energy to explain. If the activation energy of a reaction is high, the reaction rate is low. The activation energy can be lowered by catalysts, so the rate of a reaction is increased. Note that the catalyst has the same effect on the reverse reaction. So if a catalyst doubles the rate in the forward reaction, it also doubles the rate in the reverse reaction.

New Words

catalysis [kəˈtæləsɪs] *n.* 催化作用
catalyst [ˈkætəlɪst] *n.* 催化剂
increase [ɪnˈkriːs] *n.* 增加；*v.* 增加
hydrolysis [haɪˈdrɒlɪsɪs] *n.* 水解

ester ['estə(r)] n. 酯
ion ['aɪən] n. 离子
explain [ɪk'spleɪn] v. 解释，说明
note [nəʊt] vt. 注意
double ['dʌbl] vt. 使加倍

Expressions and Technical Terms

the rate of a reaction 反应速率
used up 用完，消耗完
activation energy 活化能
reverse reaction 逆反应
forward reaction 正反应

Exercises

A. Translate the following into English.
1. 催化作用　　　　2. 催化剂　　　　3. 反应速率
4. 活化能　　　　　5. 水解　　　　　6. 酯
7. 氢氧化物　　　　8. 正反应　　　　9. 逆反应

B. Decide whether the following statements are true（T）or false（F）. Write T for true and F for false in each blank.
（　　）1. A catalyst can increase the rate of a reaction, and it is used up itself in the reaction.
（　　）2. If the activation energy of a reaction is high, the reaction rate is high.
（　　）3. The activation energy can be lowered by catalysts, so the rate of a reaction is increased.
（　　）4. The catalyst has the same effect on the reverse reaction.
（　　）5. In the hydrolysis of an ester, hydroxide ion is not required in the reaction, but its addition increases the reaction rate.

C. Answer the following questions.
1. In the hydrolysis of an ester, what is the catalyst?
2. How do catalysts increase reaction rate?
3. If a catalyst doubles the rate in the forward reaction, how does it affect the rate in the reverse reaction?

Lesson 13
Reversible Reactions 可逆反应

Warming up

可逆反应： 在同一条件下，既能向正反应方向进行，同时又能向逆反应方向进行的反应。

绝大部分的反应都存在可逆性，一些反应在一般条件下并非可逆反应，而改变条件（如将反应物置于密闭环境中、高温反应等等）会变成可逆反应。

特点：

1. 反应不能进行到底。可逆反应无论进行多长时间，反应物都不可能 100% 地全部转化为生成物。

2. 可逆反应一定是同一条件下能互相转换的反应，如：二氧化硫、氧气在催化剂、加热的条件下，生成三氧化硫；而三氧化硫在同样的条件下可分解为二氧化硫和氧气。

3. 在同一时间发生的反应。

4. 同增同减。

5. 书写可逆反应的化学方程式时，应用双箭头表示，箭头两边的物质互为反应物、生成物。通常将从左向右的反应称为正反应，从右向左的反应称为逆反应。

6. 可逆反应中的两个化学反应，在相同条件下同时向相反方向进行，两个化学反应构成一个对立的统一体。在不同条件下能向相反方向进行的两个化学反应不能称为可逆反应。

7. 在外界条件不变的前提下，可逆反应不论采取何种途径，即不论由正反应开始还是由逆反应开始，最后所处的平衡状态是相同的，即同一平衡状态。

Text

Many reactions are reversible. In order to carry out a reversible reaction efficiently, the important conditions are that the reactants should be converted into products to the upmost extent and that this conversion should be brought about in the shortest possible time. For example, consider the reaction:

$$N_2 + 3H_2 \rightleftharpoons 2NH_3 + heat$$

In order to obtain a large yield of ammonia from a given quantity of nitrogen and hydrogen, the left-to-right reaction must be as nearly complete as possible. It has been found by experiment that a large yield of ammonia can be obtained under a high pressure, at a reasonable temperature and in the presence of a catalyst.

Heat is given off in the reaction, so the reaction will become more nearly complete at comparatively low temperatures. Yet the trouble is that an extremely low temperature will lower the reaction rate. Therefore, we must select a reasonable temperature. In addition to it, we must use a catalyst to speed up the left-to-right reaction.

New Words

reversible [rɪˈvɜːsəb(ə)l] *adj.* 可逆的
efficient [ɪˈfɪʃ(ə)nt] *adj.* 有效率的，（直接）生效的，能干的
upmost [ˈʌpməʊst] *adj.* 最高的，最上的
yield [jiːld] *v.* 屈服，投降，生产，获利
reasonable [ˈriːznəbl] *adj.* 合理的，有道理的

Expressions and Technical Terms

carry out 实现，完成，执行
in the shortest possible time 在尽可能短的时间内
a given quantity of... 一定量的……
under a high pressure 在高压下
at a reasonable temperature 在合适的温度下
in the presence of... 在……存在时
speed up 加速

Exercises

A. Translate the following into English.

1. 可逆反应   2. 执行   3. 在尽可能短的时间内

4. 一定量的……   5. 在高压下   6. 在合适的温度下

7. 加速   8. 在……存在时

B. Choose the best answer to each question.

1. In the production of ammonia, we must use a catalyst because (   )
（A）it can speed up the left-to-right reaction.
（B）it can lower the reaction temperature.

2. In order to obtain a large yield of ammonia, we must select (   )
（A）a very high temperature.
（B）a reasonable temperature.
（C）a very low temperature.

C. Translate the following sentences into Chinese.

1. It has been found by experiment that a large yield of ammonia can be obtained under a high pressure, at a reasonable temperature and in the presence of a catalyst.

2. Heat is given off in the reaction, so the reaction will become more nearly complete at comparatively low temperatures.

Lesson 14
Factors Affecting Reaction Rates
影响反应速率的因素

Warming up

影响反应速率的因素

一、内部因素（主要因素）

参加反应物质的性质。化学反应的本质：反应物分子中的化学键断裂、生成物分子中的化学键形成的过程（旧键断裂，新键生成）。

二、外部因素

1. 浓度：在其他条件不变时，增大反应物浓度，可以增大反应速率。

注意：a. 此规律只适用于气体或溶液的反应，对于纯固体或液体的反应物，一般情况下其浓度是常数，因此改变它们的量一般不会改变化学反应速率；b. 一般来说，固体反应物表面积越大，反应速率越大，反之，反应速率越小；c. 随着化学反应的进行，反应物的浓度会逐渐减小，因此一般反应速率也会逐渐减小。

2. 压强：对气体来说，若其他条件不变，增大压强，就是增加单位体积的反应物的物质的量，即增加反应物的浓度，因而可以增大化学反应的速率。

规律：对于有气体参加的反应，若其他条件不变，增大压强，气体体积缩小，浓度增大，反应速率加快；减小压强，反应速率减慢。

3. 温度：反应若是可逆反应，升高温度，正、逆反应速率都加快，降低温度，正、逆反应速率都减小。

4. 催化剂：使用催化剂，能够降低反应所需的能量，这样会使更多的反应物分子成为活化分子，大大增加单位体积内反应物分子中活化分子所占的百分比，因而使反应速率加快。

注意：（1）使用催化剂同等程度地增大（减慢）正逆反应速率，从而改变反应到达平衡所需时间；（2）没特别指明一般指正催化剂；（3）催化剂具有一定的选择性。

5. 其他因素：光、固体表面积、溶剂、电磁波、超声波、强磁场、高速研磨、原电池等。

Text

Concentration

Increasing the concentration increases the number of particles. Increasing the number of particles increases the number of collisions. Increasing the number of collisions increases the number of successful collisions. This increases the rate of reaction.

Pressure

Increasing the pressure means that the gas molecules are squashed into smaller volume. The same amount of gas in a smaller volume has a greater concentration. There will be more collisions, so there will be more successful collisions, so the rate will increase.

Temperature

Not all collisions between reactants succeed in making products. Only those collisions with enough energy to break bonds in the reactants will lead to a reaction. The energy is called the activation energy. Increasing the temperature of the reaction means more particles have the activation energy.

Catalyst

A catalyst allows the reaction to go by a different pathway with a lower activation energy. More particles will have this lower activation energy, and so more collisions will be successful. More successful collisions mean a higher rate, e.g. iron is added as catalyst in the Harber process for making ammonia.

New Words

factor ['fæktər] n. 因素
affect [ə'fekt] vt. 影响，假装，感动，（疾病）侵袭
concentration [ˌkɒns(ə)n'treɪʃ(ə)n] n. 集中，专心，关注，浓度
particle ['pɑːrtɪkl] n. 微粒，颗粒，[数，物] 粒子，质点，极小量
collision [kə'lɪʒ(ə)n] n. 碰撞，冲突，（意见，看法）的抵触
pressure ['preʃər] n. 压力
volume ['vɒljuːm] n. 量，大量，体积，音量，卷
catalyst ['kætəlɪst] n. 催化剂
pathway ['pæθweɪ] n. 路，道，途径，路径
iron ['aɪən] n. 铁器，铁制品
process ['prəʊses] n. 过程，工序，做事方法，工艺流程
ammonia [ə'məʊnɪə] n. 氨，氨水，氨气

Expressions and Technical Terms

reaction rate 反应速率
squash... into... 将……挤压进……
the same amount of... 等量的……
break bond 打破键
lead to 导致
activation energy 活化能

Exercises

A. Translate the following into English.

1. 因素　　　　2. 微粒　　　　3. 碰撞　　　　4. 温度

5. 压力　　　　6. 催化剂　　　7. 浓度　　　　8. 活化能

B. Decide whether the following statements are true（T）or false（F）. Write T for true and F for false in each blank.

（　）1. All collisions between reactants succeed in making products.

（　）2. Increasing the number of collisions increases the number of successful collisions. This increases the rate of reaction.

（　）3. The same amount of gas in a smaller volume has a greater concentration.

（　）4. Increasing the temperature of the reaction means more particles have the activation energy.

（　）5. A catalyst allows the reaction to go by a different pathway with a lower activation energy.

C. Translate the following sentences into Chinese.

1. Only those collisions with enough energy to break bonds in the reactants will lead to a reaction. The energy is called the activation energy.

2. A catalyst allows the reaction to go by a different pathway with a lower activation energy.

译　文

影响反应速率的因素

浓度

浓度增加，反应物的粒子数增加。粒子数增加则粒子之间的碰撞数增加。碰撞次数增加，则有效碰撞次数增加。这样反应速率就加大。

压力

压力增加表示气体分子被挤压进一个更小的空间。在一个较小的空间，同样数量的气体浓度就大了。这样碰撞更多，有效碰撞也更多，反应速率加大。

温度

并不是反应物之间所有的碰撞都能导致生成产物，只有那些有足够能量以打破反应物内的化学键的碰撞才能引起反应，这个能量叫作活化能。增加反应温度表示有更多的粒子具有活化能。

催化剂

催化剂能改变反应的途径，这种途径只要较低的活化能。由于更多的粒子具备了这种较低的活化能，这样，更多的碰撞就会是有效的。更多的有效碰撞表示更高的反应速率。例如，铁作为一种催化剂用在哈伯法生产氨中。

Lesson 15
Energy and Chemical Energy
能量与化学能

Warming up

势能： 储存于一个系统内的能量，也可以释放或者转化为其他形式的能量。势能是相互作用的物体所共有的能量。选择不同的势能零点，势能的数值一般是不同的。势能按作用性质的不同，可分为引力势能、弹性势能、电势能等。

动能： 物体由于运动而具有的能量。它的大小定义为物体质量与速度平方乘积的二分之一。物体的运动速度越大，物体的质量越大，物体的动能就越大。

化学能： 是一种很隐蔽的能量，它不能直接用来做功，只有在发生化学变化的时候才可以释放出来，变成热能或者其他形式的能量。各种物质都储存有化学能。不同的物质不仅组成不同、结构不同，所包含的化学能也不同。在化学反应中，既有反应物中化学键的断裂，又有生成物中化学键的形成，那么，一个确定的化学反应完成后的结果是吸收能量还是放出能量，决定于反应物的总能量与生成物的总能量的相对大小。

化学反应在反应前后遵循质量守恒定律；同样，一种能量可以转化为另一种能量，能量也是守恒的，这就是"能量守恒定律"。

化学反应伴随着能量变化是化学反应的一大特征。化学物质中的化学能通过化学反应转化成热能，是人类生存和发展的动力之源；而热能转化为化学能又是人们进行化学科学研究，创造新物质不可缺少的条件和途径。

Text

Matter can have both potential and kinetic energy. Potential energy is stored-up energy. For example, water behind a dam has potential energy and it can be changed into electrical energy. Gasoline has chemical potential energy. It can be released during combustion.

All moving bodies have kinetic energy. When the water behind the dam is allowed to flow, its potential energy is changed into kinetic energy. The kinetic energy can be used to produce electricity.

The common kinds of energy are mechanical, chemical, heat, electrical and light energy.

In all chemical reactions, matter either absorbs or releases energy. Chemical reactions can be used to produce different kinds of energy. For example, heat and light energies are released from the combustion of fuels.

Conversely, energy can be used to cause chemical reactions. For example, a chemical reaction occurs when light energy is used by plants. A chemical reaction also occurs when electrical energy causes water to decompose.

One type of energy may be changed into energy of another type. For example, when hydrogen and oxygen are burned, chemical energy is changed into light and heat energy. When electrical energy decomposes water again producing hydrogen and oxygen, electrical energy is changed into chemical energy.

New Words

potential [pəˈtenʃ(ə)l] *adj*. 潜在的，可能的
kinetic [kɪˈnetɪk] *adj*. 运动的，动的
dam [dæm] *n*. 水坝
gasoline [ˈgæsəliːn] *n*. 汽油
combustion [kəmˈbʌstʃən] *n*. 燃烧
absorb [əbˈzɔːb] *vt*. 吸收，吸引
fuel [ˈfjuːəl] *n*. 燃料
conversely [ˈkɒnvɜːslɪ] *adv*. 相反地
cause [kɔːz] *n*. 原因；*vt*. 引起
occur [əˈkɜː(r)] *vi*. 发生，出现

Expressions and Technical Terms

potential energy 势能
kinetic energy 动能
stored-up 储存
be changed into… 被转变成……
electrical energy 电能
mechanical energy 机械能
chemical energy 化学能
heat energy 热能
light energy 光能

Exercises

A. Translate the following into English.

1. 动能　　　　2. 势能　　　　3. 电能

4. 机械能　　　5. 化学能　　　6. 热能

7. 光能　　　　8. 吸收能力　　9. 释放能量

B. Decide whether the following statements are true (T) or false (F). Write T for true and F for false in each blank.

(   ) 1. Water behind a dam has potential energy and it can be changed into electrical energy.

(   ) 2. All moving bodies have kinetic energy.

(   ) 3. The common kinds of energy are mechanical, chemical, heat, electrical and light energy.

(   ) 4. In all chemical reactions, matter neither absorbs nor releases energy.

(   ) 5. One type of energy may be changed into energy of another type.

C. Fill in the blanks.

1. Matter can have both _____ and _____ energy.
2. All moving bodies have _____ .
3. The common kinds of energy are _____ energy.
4. In all chemical reactions, matter either _____ or _____ energy.
5. Energy can be used to cause _____ .
6. One type of energy may be _____ energy of another type.

Part 2

Health, Safety, Environment (HSE)

健康、安全和环境

Lesson 16
Hazards in Chemical Engineering Laboratories
化学工程实验室中的危险

Warming up

学生实验守则

为了顺利完成实验任务，确保人身、设备安全，培养学生诚信、严谨、安全、环保的科学作风和爱护国家财产的优秀品质，要求每个学生必须遵守实验制度：

1. 实验前应充分预习，认真阅读实验指导书，明确实验任务及要求，弄清实验及仪器原理。 拟定好实验方案，教师应对预习情况进行检查。

2. 使用仪器设备前，必须熟悉其性能，预习操作方法及注意事项，使用时严格遵守操作规程，做到准确操作。 非本次实验所用仪器设备未经教师允许不得动用。

3. 实验中应注意观察现象，认真记录实验结果。 实验记录经老师审阅检查，签字登记。

4. 遵守实验纪律，注意保持实验室整洁、安静，不做与实验内容无关的事情。

5. 实验结束后，关掉仪器设备电源开关，将仪器设备整理好，实验现场整理干净，方可离开实验室。

6. 实验过程中如发生事故，要服从指导教师对事故的处理。 事后应自觉写事故报告，总结经验，吸取教训。 借用器材有损坏、丢失，要按实验室规定赔偿。

7. 实验完成后，应按要求认真书写实验报告，按时交给老师。 老师根据实验记录与实验报告情况进行评分。

Text

For most students, working in a chemical engineering laboratory introduces them to operate on a much larger scale. Other changes are the increased quantity of material involved, the increased energy in systems, and the added complexity of dealing with runaway reactions.

Flammable liquids are used as reactants, solvents, or cleaning fluids. While safety may be improved by substitution, the strictest vigilance must be maintained on toxicity hazards. If flammable liquids are employed, the practical and physical significance of their properties should be explained to the users. The hazard should be minimized by limiting the quantity in the working area, providing adequate licensed storage and instituting a workable system of handling and withdrawal from the store. Sources of ignition must be eliminated. Coordination with other work in the vicinity is also important.

Students are unlikely to work with solid explosives but may become involved with oxidizing agents and combustible solids. These demand respect in handling and storage. The same applies to exothermic liquid phase reactions and "sealed vessel" reactions. To avoid gas and vapor explosions, it is necessary to avoid the confluence of flammable mixtures and an ignition source.

The student is likely to encounter several new aspects of the hazards of electricity: many are unfamiliar with the wiring of three-phase appliances and with the safe use of separate phases. All such wiring must be the responsibility of a trained electrician. High-powered immersion heaters require adequate cutout devices, fail-to-safety devices and earth leakage protection should be more widely used. Static electrical charges can present a hazard of ignition, shock or nuisance. Therefore adequate earthing or neutralization of charge should be incorporated into the apparatus.

Toxic materials are often used. If necessary, however, complete or partial containments should be used. Attention must then be directed to the quality and fate of any exhaust stream. Ventilation is a requirement under almost all circumstances. The hazard of sudden releases must be quantitatively assessed and large stocks of toxic material must be kept outside the laboratory.

It is concluded that hazards should be identified at the design stage by supervisor. Hazards can then be engineered-out. The effect of any hazard which cannot be engineered-out must be quantified. This should be part of the normal training of any worker.

New Words

hazard [ˈhæzəd] *n*. 危险，冒险的事
engineer [ˌendʒɪˈnɪə(r)] *n*. 工程师，技师
laboratory [ˈlæbrətɔːrɪ] *n*. 实验室，实验课，研究室，药厂
quantity [ˈkwɒntətɪ] *n*. 量，数量，定量
complexity [kəmˈpleksətɪ] *n*. 复杂性，错综复杂的状态
runaway [ˈrʌnəweɪ] *adj*. 逃走的，失去控制的
substitution [ˌsʌbstɪˈtjuːʃn] *n*. 代替，替换，取代（作用）
vigilance [ˈvɪdʒɪləns] *n*. 警觉，警惕，警戒
toxicity [tɒkˈsɪsətɪ] *n*. 毒性，毒力
practical [ˈpræktɪkl] *adj*. 实践的，实际的，实用的
significance [sɪɡˈnɪfɪkəns] *n*. 重要性，意义，意思
adequate [ˈædɪkwət] *adj*. 足够的，适当的
license [ˈlaɪsns] *vt*. 同意，发许可证
institute [ˈɪnstɪtjuːt] *vt*. 建立，制定
withdrawal [wɪðˈdrɔːəl] *n*. 移开，撤回，撤退
ignition [ɪɡˈnɪʃn] *n*. 发火装置，燃烧，点火
eliminate [ɪˈlɪmɪneɪt] *vt*. 排除，消除，淘汰
coordination [kəʊˌɔːdɪˈneɪʃn] *n*. 协调，和谐

vicinity [vəˈsɪnətɪ] adj. 附近，邻近，附近地区
explosive [ɪkˈspləʊsɪv] adj. 爆炸的，易爆炸的；n. 爆炸物，炸药
exothermic [ˌeksəʊˈθɜːmɪk] adj. 发热的，放出热量的
confluence [ˈkɒnfluəns] n. 汇合，汇流
appliance [əˈplaɪəns] n. 器具，装置，家用电器
responsibility [rɪˌspɒnsəˈbɪlətɪ] n. 责任，职责，责任感，负责任
immersion [ɪˈmɜːʃn] n. 沉浸，洗礼
nuisance [ˈnjuːsns] n. 非法妨害，损害，麻烦事
apparatus [ˌæpəˈreɪtəs] n. 仪器，器械，机器
fate [feɪt] n. 宿命，灾难
ventilation [ˌventɪˈleɪʃn] n. 空气流通，通风设备
circumstance [ˈsɜːkəmstəns] n. 环境，境遇
supervisor [ˈsuːpəvaɪzə(r)] n. 管理者，监督者，指导者

Expressions and Technical Terms

sealed vessel 密封容器
cutout devices 切断装置
fail-to-safety devices 故障安全装置
earth leakage protection 对地漏电保护
static electrical charges 静电电荷
exhaust stream 排气流

Exercises

A. Translate the following into English.
1. 危险   2. 化工实验室   3. 可燃液体   4. 毒性

5. 点火装置   6. 易爆炸的   7. 切断装置   8. 警觉

B. Decide whether the following statements are true (T) or false (F). Write T for true and F for false in each blank.
(　　) 1. Toxic materials are often used.
(　　) 2. Flammable liquids are used as reactants, solvents, or cleaning fluids.
(　　) 3. The hazard should be minimized by limiting the quantity in the working area.
(　　) 4. To avoid gas and vapor explosions, it is necessary to avoid the confluence of flammable mixtures and an ignition source.
(　　) 5. The student is unlikely to encounter several new aspects of the hazards of electricity.

C. Choose the best answer to each question.
1. A student working in a chemical engineering laboratory is expected to (　　)
(A) have rather limited experience.

(B) see strange experiments.
(C) know operations on a larger scale.
(D) deal with runaway reactions only.

2. When a student deals with flammable liquids in a lab he must (　　)
(A) keep away from toxicity hazards.
(B) maintain little vigilance.
(C) remain sleepless.
(D) know the practical and physical significance of their properties.

3. We are told that flammable liquids can be used as all the following except (　　)
(A) cleaning fluids.　　(B) pills.　　(C) reactants.　　(D) solvents.

4. Which of the following materials is a student most unlikely to handle in a lab? (　　)
(A) combustible solids.　　(B) flammable liquids.
(C) solid explosives.　　(D) oxidizing agents.

5. In a lab, all the wiring should be handled by (　　)
(A) a student.　　(B) a group of students.
(C) an electrician.　　(D) a senior student.

6. Where toxic gases are used, there should be (　　)
(A) at least two students.　　(B) an electrician.
(C) sealed vessels.　　(D) monitoring.

译　文

化学工程实验室中的危险

对大部分学生而言，在化工实验室工作使他的操作范围大大增加。其他改变还包括接触到更大量的物质，系统中更高的能量，以及在处理失控反应时将面对更强的复杂性。

易燃液体可用作反应物、溶剂或清洗流体。虽然可以通过更换（易燃液体的品种）来提高安全性，但对于危险化学品的危险性仍须保持极高的警惕。如果使用了易燃液体，那么它们性质上的实际意义和物理意义一定要给使用者说明。如果在工作场所限制（易燃物品的）存放量，同时提供足够合格的储藏室，以及制定切实可行的处理和领用易燃物品的制度，那么危险性就可减少到最低程度。火源必须消除，协调邻近区域的其他工作也很重要。

学生不可能用到固体爆炸物但可能用到氧化剂和易燃固体，这些物质的处理和储藏必须谨慎。同样的原理也适用于放热的液相反应和密闭容器内的反应。为了避免气体和蒸气爆炸，必须防止易燃混合物和火源的接触。

学生可能会遇到用电方面的危险：学生不熟悉三相装置的布线，也不熟悉每一相的安全使用。所有这些布线必须是受过训练的电工的职责，高功率的浸入式加热器需要合适的切断装置，故障安全保险装置和接地保护装置应该被更广泛地使用。静电荷将引起着火、电击或其他损害，因此设备中应有适当的接地或中和电荷的装置。

有毒物质经常会被使用。然而，如果必要的话，应对它们完全或部分限制使用。必须注意所有废气的量和流动情况，几乎所有的环境都要求通风。毒物突然释放的危险必须进行定量的评价，大量有毒的物质必须排除在实验室外。

可以得出结论，危险应由监管人员在设计阶段就示出。然后，危险才能被巧妙地排除，对于任何无法排除的危险所引起的后果必须量化。这应该是所有工作人员常规训练的必要内容。

Lesson 17
Environment Protection
环境保护

Warming up

环境保护一般是指人类为解决现实或潜在的环境问题，协调人类与环境的关系，保护人类的生存环境、保障经济社会的可持续发展而采取的各种行动的总称。

环境保护方式包括：采取行政、法律、经济、科学技术等手段和方法，合理利用自然资源，防止环境的污染和破坏，以求自然环境同人文环境、经济环境共同平衡可持续发展，扩大有用资源的再生产，保证社会的发展。

随着人类对环境认识的深入，环境是资源的观点越来越为人们所接受。空气、水、土壤、矿产资源等，都是社会的自然财富和发展生产的物质基础，构成了生产力的要素。

联合国环境规划署标志

Text

Some Important Definitions

(1) Environment is the physical and biotic habital which surrounds us, that which we can see, hear, touch, smell, and taste.

(2) System is defined as a set or arrangement of things so related or connected as to form a unit or organic whole, as a solar system, irrigation system, supply system, the world or universe according to Webster's dictionary.

(3) Pollution can be defined as an undesirable change in the physical, chemical, or biological characteristics of the air, water, or land that can harmfully affect the health, survival, or activities of humans or other living organisms.

Manufacturing with Minimal Environmental Impact

Increasing concern over adverse health effects has put a high priority on eliminating or reducing the amounts of potentially hazardous chemicals used in industrial processes. The best course of action is to find replacement chemicals that work as well but are less hazardous. Recent advances in catalysis, for example, permit chemical reactions to be run at lower temperature and pressures. This change, in turn, reduces the energy demands of the process and simplifies the selection of construction materials for the processing facility.

Environmentally Friendly Products

Increasing understanding of the fate of products in the environment has led scientists to design "greener" products. Novel biochemistry is also helping farmers reduce the use of insecticides. Cotton plants, for example, are being genetically modified to make them resistant to the cotton bollworm.

Recycling

Increasing problems associated with waste disposal have combined with the recognition that some raw materials exist in limited supply to dramatically increase interest in recycling. Recycling of metals and most paper is technically straightforward, and these materials are now commonly recycled in many areas around the world. Recycling of plastics presents greater technical challenges.

Separation and Conversion for Waste Reduction

New processes are needed to separate waste components requiring special disposal from those that can be recycled or disposed by normal means, such as waste treatment, membrane technology, and biotechnology.

New Words

    environment [ɪnˈvaɪrənmənt] *n.* 生活环境，自然环境，生态环境
    definition [ˌdefɪˈnɪʃn] *n.* 定义，释义
    biotic [baɪˈɒtɪk] *adj.* 关于生命的，生物的
    habital [ˈhæbɪtl] *n.* 生物环境
    system [ˈsɪstəm] *n.* 体系，制度，系统
    arrangement [əˈreɪndʒmənt] *n.* 安排，筹划，布置，商定，协议，整理
    irrigation [ˌɪrɪˈgeɪʃn] *n.* 灌溉，冲洗
    undesirable [ˌʌndɪˈzaɪərəbl] *adj.* 不受欢迎的，不良的
    survival [səˈvaɪv(ə)l] *n.* 生存，残存物
    organism [ˈɔːgənɪzəm] *n.* 有机体，生物体，有机体系，有机组织
    manufacture [ˌmænjuˈfæktʃə(r)] *vt.* 制造，生产，加工，从事制造
    priority [praɪˈɒrəti] *n.* 优先，优先权，重点
    facility [fəˈsɪləti] *n.* 设施
    fate [feɪt] *n.* 命运，注定，宿命
    insecticide [ɪnˈsektɪsaɪd] *n.* 杀虫剂
    resistant [rɪˈzɪstənt] *adj.* 有抵抗力的
    bollworm [ˈbəʊlwɜːm] *n.* 一种蛾的幼虫，螟蛉
    disposal [dɪˈspəʊzl] *n.* 处理，清除
    recognition [ˌrekəgˈnɪʃn] *n.* 认识，识别
    dramatically [drəˈmætɪklɪ] *adv.* 戏剧地，显著地
    straightforward [ˌstreɪtˈfɔːwəd] *adj.* 简单明了的，坦率的

plastic [ˈplæstɪk] n. 塑料制品

Expressions and Technical Terms

environment protection 环境保护
solar system 太阳系
irrigation system 灌溉系统
supply system 供应系统
minimal environmental impact 最小的环境影响
hazardous chemicals 危险化学品
replacement chemicals 替代化学品
energy demands 能源需求
construction materials 建筑材料
raw materials 原材料
metal-bearing spent acid waste 含金属的废酸
high-level nuclear waste 高含量的核废料
membrane technology 膜技术

Exercises

A. Translate the following into English.

1. 环境保护   2. 杀虫剂   3. 灌溉系统   4. 供应系统

5. 最小的环境影响   6. 危险化学品   7. 能源需求   8. 建筑材料

9. 核废料   10. 膜技术   11. 替代化学品

B. Decide whether the following statements are true (T) or false (F). Write T for true and F for false in each blank.

(   ) 1. Environment is that which we can see, hear, touch, smell, and taste.

(   ) 2. Increased understanding of the fate of products in the environment has led scientists to design "greener" products.

(   ) 3. Pollution can harmfully affect the health, survival, or activities of humans or other living organisms.

(   ) 4. The best course of action is to find replacement chemicals that work as well but are less hazardous.

(   ) 5. Recycling of metals and most paper is technically straightforward, and these materials are now commonly recycled in many areas around the world.

C. Fill in the blanks.

1. Environment is the _____ and _____ habital which surrounds us.
2. Systemis defined as a set or arrangement of things so related or connected as to form a unit

or _____ whole.

3. Pollution is an undesirable _____ in the physical, chemical, or biological characteristics of the air, water, or land that can harmfully affect the _____, survival, or activities of humans or other _____ organisms.

4. Recycling of _____ presents greater technical challenges.

5. New processes are needed to separate waste _____ requiring special disposal.

Reading Material 阅读材料

环境保护

1. 几个重要的定义

环境是我们周围可以看到、听到、触到、嗅到和尝到的物理和生物环境。

系统根据韦氏词典被定义为一系列相关联事件组合而形成的一个单元或一个有机整体，像太阳系、灌溉系统、供应系统、世界或宇宙。

污染为空气、水、土地在物理、化学或生物特性上的不良变化，这些变化严重影响人类的健康、生存、活动或者其他生命有机体。

当改善环境是为了提高人类生活的幸福度时，环境这个词拓宽到社会、经济、文化等各个方面。在课堂里设置一个学期时长的学习达不到环境保护真正涵盖内容的广度。

2. 对环境影响最小的生产

随着人们对健康危害的担忧日益增加，使得在工业过程中，消除或减少潜在有害化学品的使用被放到优先考虑的位置。最佳的方案是找到替代的化学品，其效果良好且无害。如果不能找到有害化学物的替代品，一个有战略前景的选择是开发一种在现场生产、且仅生产当时所需消耗量的工艺。

化学的创新已开始提供有益于环境的合理工艺，其能更加有效地使用能源和原材料。例如，在催化剂方面，近期的进展就使化学反应在较低的温度和压力下进行。反过来，这种变化能降低生产过程中的能源需求，并简化加工设备对建筑材料的选择。新颖的催化剂也被用来避免生产不必要的副产品。

3. 环保产品

增加对环境中产品归宿的理解促使科学家设计"更加环保"的产品。一个值得注意的早期例子来自20世纪的40和50年代洗涤剂产业；新产品是在合成接枝的烷基苯磺酸盐这类表面活性剂的基础上引进的。这种去污剂清洗效率更高，但是结果却发现，它们在废水中的存在让溪流与河水产生泡沫。问题被追溯到接枝的烷基苯磺酸盐；这不像先前所使用的香皂，在常规污水处理厂中这些盐不能被微生物充分地生物降解。化学家们努力了解适当的生物化学过程并进行大量的研究，设计和合成另一类新的表面活性剂：线型烷基苯磺酸盐。这些新化合物和传统肥皂中天然的脂肪酸在分子结构之间的相似性，让微生物能降解这些新的制剂，而且这种相似也让接枝的烷基苯磺酸盐表现显著的去污性能。

新生物化学也帮助农民减少对杀虫剂的使用。例如，通过转基因使棉花植株能抵抗棉铃虫。当天然细菌里的单基因转育到棉花上时，这种基因促使棉花自身产生一种杀虫蛋白质。当棉铃虫吞食棉花时，蛋白质通过阻断其消化过程将其杀死。

4. 回收

越来越多的问题与垃圾处理有关。人们已经达成共识：一些原材料存量有限，这加剧了人们对废品回收的兴趣。回收金属和大多数废纸的技术很简单，而且现在世界上许多地区已实现了对金属和废纸的回收，但塑料回收的技术难度较大。

即使它们与其他废弃物分离后，不同的塑料材料也必须彼此分离。在这种情况下，不同类型的塑料有不同的化学性质，就需要开发不同的回收工艺。

一些塑料可以通过简单的熔化和成形回收或是通过溶解在适当的溶剂中形成一种新的塑料材料。另外一些材料需要更复杂的处理，如把大的聚合物分子分解成小的子单元，随后被用作构建新的聚合物。事实上，一个通过该路线回收塑料饮料瓶的主要项目现在正在使用。

化学家和化学工程师需要大量研究工作才能成功开发所需要的回收技术。在这样的情况下，很有必要开发全新分子结构的聚合物以便于回收。

5. 为减少废弃物的分离和转化

需要新的工艺来分离废弃物，从那些可以通过常规方法回收或处理的废弃物中再一次进行回收处理。这些工艺的发展需要广泛的研究，以获得对有关化学现象基本的理解。

含金属的废酸废弃物　有些工业过程会产生大量的酸性废液。这种废液能够被分离成干净的水、可重复使用的酸和重新获得金属的污泥。这些工艺会保护环境，并且它们的成本与处理废物成本和相应的处罚费用相当。

废物处理　热催化或光催化过程可能破坏工业废水中有害的有机成分。在高温和高压下运用"超临界"水是一种很有前景的研究路线。在这种情况下，水呈现出不同寻常的物理和化学性质。它能溶解在正常条件下几乎是惰性的物质，并允许其中的许多物质反应。

核废物　通过减少核废料储存的成交量和补偿量，可节省巨额费用；这样的减少需要对含核废弃物的大量物质进行放射性成分的有效分离。有害的化学废弃物可被分别处理掉。核废弃物处理需要经过多年的专业研究和开发投入。

膜分离技术　涉及半透膜的分离展现出一个相当好的前景。这种通常是聚合物的薄膜，一些化学物质不能渗透，而另一些能渗透。这种薄膜可用于净化水，能除去水中溶解的盐类，提供干净的饮用水。膜分离技术也用于生产中净化废水。膜分离也适用于气体，被用于去除天然气中的微量杂质成分，通过清除二氧化碳提高天然气的燃烧值，并对空气中的氮进行回收。该研究的挑战包括新膜的开发，来改善膜的化学和物理性质，使其更富弹性，且降低生产成本，并能够提供更好的分离效率来降低加工成本。

生物技术　科学家已经向大自然寻求帮助来消除有毒的物质。在土壤、水和沉积物中的微生物把各种各样的有机化学物质作为食物；它们在废物处理体系中的应用已有几十年的历史。通过仔细地测定这些"有天赋的"微生物存在的最佳物理、化学和营养条件，研究人员试图催生更高性能的微生物。他们的努力可能促进新一代的生物技术废品处理设备的设计及应用。近年一个重大进展是这些微生物在生物反应器中的固定，当它们降解废弃物时会被固定在一个反应器中。微生物固定在传统反应器中能实现高流速冲刷，并能使用新型的、更多孔的支撑物质，这样就可以使每个反应器中微生物数量有显著的增长。

Lesson 18
Green Chemistry
绿色化学

Warming up

绿色化学又称环境无害化学、环境友好化学、清洁化学，是用化学的技术和方法减少或消除有害物的生产和使用。绿色化学的定义是在不断地发展和变化的。

绿色化学涉及有机合成、催化、生物化学、分析化学等学科，内容广泛。绿色化学倡导用化学的技术和方法减少或停止那些对人类健康、社区安全、生态环境有害的原料、催化剂、溶剂和试剂、产物、副产物等的使用与产生。

绿色化学与污染控制化学不同。绿色化学的设想是使污染消除在产生的源头，使整个合成过程和生产过程对环境友好，不再使用有毒、有害的物质，不再产生废物，不再处理废物，这是从根本上消除污染的对策。由于在始端就采用预防污染的科学手段，过程和终端均为零排放或零污染。世界上很多国家已把"化学的绿色化"作为新世纪化学进展的主要方向之一。

绿色化学利用可持续发展的方法。近年来，绿色化学的研究主要围绕化学反应、原料、催化剂、溶剂和产品的绿色化来进行。我国科研学者提出了绿色化工产品设计、原料绿色化及新型原料平台、新型反应技术、催化剂制备的绿色化和新型催化技术、溶剂的绿色化及绿色溶剂、新型反应器及过程强化与耦合技术、新型分离技术、绿色化工过程系统集成、计算化学与绿色化学化工结合等9个方面绿色化学和化工的发展趋势。

Text

Green chemistry, also called sustainable chemistry, is a philosophy of chemical research and engineering that encourages the design of products and processes that minimize the use and generation of hazardous substances. Whereas, environmental chemistry is the chemistry of the natural environment, and of pollutant chemicals in nature. Green chemistry seeks to reduce and prevent pollution at its source.

As a chemical philosophy, green chemistry applies to organic chemistry, inorganic chemistry, biochemistry, analytical chemistry, and even physical chemistry. While green chemistry seems to focus on industrial applications, it does apply to any chemistry choice. Click chemistry is often cited as a style of chemical synthesis that is consistent with the goal of green chemistry. The focus is on minimizing the hazard and maximizing the efficiency of any chemical choice.

Green chemistry protects the environment, not by cleaning up, but by inventing new chemistry and new chemical processes that do not pollute. Green chemistry emphasizes renewable starting materials for a bio-based economy.

Twelve Principles of Green Chemistry

(1) Prevention

It is better to prevent waste than to treat waste after it has been created.

(2) Atom economy

Synthetic methods should be designed to maximize the incorporation of all materials used in the process into the final product.

(3) Less hazardous chemical syntheses

Wherever practicable, synthetic methods should be designed to use and generate substances that possess little or no toxicity to human health and the environment.

(4) Designing safer chemicals

Chemical products should be designed to effect their desired function while minimizing their toxicity.

(5) Safer solvents and auxiliaries

The use of auxiliary substances (e.g., solvent, separation agents, etc.) should be made unnecessary wherever possible and innocuous when used.

(6) Design for energy efficiency

Energy requirements of chemical processes should be recognized for their environmental and economic impacts and should be minimized. If possible, synthetic methods should be conducted at ambient temperature and pressure.

(7) Use of renewable feedstocks

A raw material or feedstock should be renewable rather than depleting whenever technically and economically practicable.

(8) Reduce derivatives

Unnecessary derivatization (use of blocking groups, protection/deprotection, temporary modification of physical/chemical processes) should be minimized or avoided if possible, because requiring additional reagents can generate waste.

(9) Catalysis

Catalytic reagents (as selective as possible) are superior to chemical reagents.

(10) Design for degradation

Chemical products should be designed at the end of their function they break down into innocuous degradation products and do not persist in the environment.

(11) Real-time analysis for pollution prevention

Analytical methodologies need to be further developed to allow for real-time, in-process monitoring and control prior to the formation of hazardous substances.

(12) Inherently safer chemistry for accident prevention

Substances and the form of a substance used in a chemical process should be chosen to minimize the potential for chemical accidents, including releases, explosions, and fires.

Green chemistry is seen as a powerful tool that researchers must use to evaluate the environmental impact of nanotechnology, as nanomaterials and the processes to make them must be considered to ensure their long-term economic viability.

New Words

sustainable [səˈsteɪnəbl] *adj*. 可持续的
philosophy [fəˈlɒsəfi] *n*. 哲学，哲学思想
encourage [ɪnˈkʌrɪdʒ] *v*. 鼓励，使有希望
minimize [ˈmɪnɪmaɪz] *vt*. 把……减至最低数量 [程度]
click [klɪk] *v*. 使发出咔嗒声，(用鼠标) 点击，受到欢迎，配合默契
cite [saɪt] *v*. 引用，引证
efficiency [ɪˈfɪʃnsi] *n*. 效率
invent [ɪnˈvent] *v*. 发明
emphasize [ˈemfəsaɪz] *v*. 强调，重视，使突出
economy [ɪˈkɒnəmi] *n*. 经济，经济制度
prevention [prɪˈvenʃn] *n*. 预防，阻止
toxicity [tɒkˈsɪsəti] *n*. 毒性，毒力
auxiliary [ɔːɡˈzɪliəri] *n*. 助动词，辅助人员
innocuous [ɪˈnɒkjuəs] *adj*. 无害的，平淡无味的
recognize [ˈrekəɡnaɪz] *vt*. 认出，识别
conduct [kənˈdʌkt] *v*. 进行，组织，实施
ambient [ˈæmbiənt] *adj*. 周围的，包围着的，环境
renewable [rɪˈnjuːəbl] *adj*. 可更新的，可再生的
feedstock [ˈfiːdstɒk] *n*. 给料，进料
deplete [dɪˈpliːt] *v*. (大量) 减少，消耗
practicable [ˈpræktɪkəbl] *adj*. 切实可行的，实际的，实用的
derivative [dɪˈrɪvətɪv] *n*. 派生物，衍生物
block [blɒk] *n*. 一块，立方体，大楼，障碍 (物)，拦截
superior [suːˈpɪəriə(r)] *adj*. 比……好的，更高级的
degradation [ˌdeɡrəˈdeɪʃn] *n*. 衰退，降解
inherent [ɪnˈherənt] *adj*. 固有的，内在的
nanotechnology [ˌnænəʊtekˈnɒlədʒi] *n*. 纳米技术
viability [ˌvaɪəˈbɪləti] *n*. 生存能力

Expressions and Technical Terms

sustainable chemistry 可持续化学
environmental chemistry 环境化学

at its source 在源头上
chemical synthesis 化学合成
atom economy 原子经济
separation agent 分离剂
energy efficiency 能量效率
temporary modification 临时性修改
catalytic reagent 催化试剂
innocuous degradation product 无害的降解产物
real-time analysis 实时分析
analytical methodology 分析方法
prior to… 在……之前
accident prevention 事故预防

Exercises

A. Translate the following into English.
1. 绿色化学   2. 可持续化学   3. 原子经济   4. 能量效率

5. 可再生原料   6. 实时分析   7. 无害的降解产物   8. 事故预防

B. Decide whether the following statements are true (T) or false (F). Write T for true and F for false in each blank.

(　) 1. Green chemistry seeks to reduce and prevent pollution at its source.

(　) 2. Green chemistry protects the environment, by cleaning up, by inventing new chemistry and new chemical processes that do not pollute.

(　) 3. Green chemistry emphasizes renewable starting materials for a bio-based economy.

(　) 4. It is worse to prevent waste than to treat waste after it has been created.

(　) 5. A raw material or feedstock should be renewable rather than depleting.

(　) 6. Catalytic reagents are superior to chemical reagents.

(　) 7. Analytical methodologies need to be further developed to allow for real-time in-process monitoring.

C. Answer the following questions.
1. What is green chemistry?
2. What is environmental chemistry?
3. What are the twelve principles of green chemistry?

Part 3
Analytical Chemistry
分析化学

Lesson 19
Analytical Chemistry and Chemical Analysis
分析化学和化学分析

Warming up

分析化学包括化学分析和仪器分析。

化学分析：指确定物质化学成分或组成的方法。根据被分析物质的性质可分为无机分析和有机分析。根据分析的要求，可分为定性分析和定量分析。根据被分析物质试样的数量，可分为常量分析、半微量分析、微量分析和超微量分析。

工业上的原材料、半成品、成品，农业上的土壤、肥料、饲料，以及交通运输上的燃料、润滑剂等，在研究、试制、生产或使用的过程中，都需要应用化学分析。

化学分析根据其方法的不同，可分为滴定分析、重量分析等。

滴定分析：根据滴定所消耗标准溶液的浓度和体积，以及被测物质与标准溶液所进行的化学反应计量关系，求出被测物质的含量，这种分析也叫容量分析。

利用溶液四大平衡：酸碱（电离）平衡、氧化还原平衡、配位平衡、沉淀溶解平衡。滴定分析根据其反应类型的不同，可将其分为下面四类。

1. 酸碱滴定法：测各类酸碱的酸碱度和酸碱的含量；
2. 氧化还原滴定法：测具有氧化还原性的物质；
3. 配位滴定法：测金属离子的含量；
4. 沉淀滴定法：测卤素和银。

重量分析：通过适当的方法如沉淀、挥发、电解等，使待测组分转化为另一种纯净的、化学组成固定的化合物而与样品中其他组分得以分离，然后称其质量，根据称得到的质量计算待测组分的含量。

重量分析法适用于待测组分含量大于 1% 的常量分析，其特点是准确度高，因此此法常被用于仲裁分析，但操作麻烦、费时。重量分析的基本操作包括：样品溶解、沉淀、过滤、洗涤、烘干和灼烧等步骤。

仪器分析：以物质的物理和物理化学性质为基础的分析方法称物理和物理化学分析法，这类方法都需要较特殊的仪器，通常称为仪器分析方法。

最主要的仪器分析方法有以下几种：

> 1. 光学分析法
> 根据物质的光学性质所建立的分析方法。主要包括：分子光谱法，分光分析法，分子荧光及磷光分析法，原子光谱法，如原子发射光谱法、原子吸收光谱法。
> 2. 电化学分析法
> 根据物质的电化学性质所建立的分析方法。主要包括电位分析法、极谱和伏安分析法、电重量和库仑分析法、电导分析法。
> 3. 色谱分析法
> 根据物质在两相（固定相和流动相）中吸附能力、分配系数或其他亲和作用的差异而建立的一种分离、测定方法。这种分析法最大的特点是集分离和测定于一体，是多组分物质高效、快速、灵敏的分析方法。主要包括气相色谱法、液相色谱法。

Text

There are many examples of how chemical tests are used to solve everyday problem. These examples include techniques for monitoring pollutants in air or water and methods for detecting bacteria or contaminants in our food, textiles, drugs, plastics, and metals. In addition, chemical analysis plays an important role in forensic science and clinical testing, and is a vital component of research in biology, biochemistry, medicine, and materials science. In fact, almost every day your life is probably affected in some way by chemical analysis.

The field of chemistry that deals with the use and development of tools and processes for examining and studying chemical substances is known as analytical chemistry.

Perhaps the most functional definition of analytical chemistry is that it is "the qualitative and quantitative characterization of matter". It may mean the identification of the chemical compounds or elements present in a sample to answer questions, such as "Is there any vitamin E in this shampoo as indicated on the label?" This type of characterization, to tell us what is present is called qualitative analysis. Qualitative analysis is the identification of one or more chemical species present in a material. Characterization may also mean the determination of how much of a particular compound or element is present in a sample, to answer questions such as "How much nickel is there in the steel?" This determination of how much of a species is present in a sample is called quantitative analysis. Quantitative analysis is the determination of the exact amount of a chemical species present in a sample. The chemical species may be an element, a compound, or an ion. The compound may be organic or inorganic.

For many years, analytical chemistry relied on chemical reactions to identify and determine the components present in a sample. These types of classical methods, often called "wet chemical methods", usually required that a part of the sample is taken, dissolved in a suitable solvent if necessary and the desired reaction is carried out. The most important analytical fields based on this approach were volumetric and gravimetric analysis. These types of analyses require a high degree of skill and attention to details on the part of the analyst if accurate and precise results are to be obtained. They are also time consuming, but many of the volumetric methods have been transferred to automated instruments. Classical analysis and instrumental analysis are similar in many respects, such as in the need for proper sampling, sample preparation, assessment of accuracy and precision, and proper record keeping.

Most analyses are carried out with specially designed electronic instruments controlled by computers. In addition, it may be necessary to analyze samples without destroying them. The analytical chemist must not only know and understand analytical chemistry and instrumentation, but must also be able to serve as a problem solver to colleagues in other scientific areas. The field of analytical chemistry is advancing rapidly. To keep up with the advances, the analytical chemist must understand materials science, metallurgy, biology, food science, geology, and other fields. The modern analytical chemists often must also consider factors such as time limitations and cost limitations in providing an analysis.

New Words

monitor ['mɒnɪtə(r)] v. 监控，监听；n. 监测仪器
pollutant [pə'luːtənt] n. 污染物
detect [dɪ'tekt] v. 发现，查明，测出
bacteria [bæk'tɪərɪə] n. 细菌（bacterium 的复数）
contaminant [kən'tæmɪnənt] n. 污染物，致污物
textile ['tekstaɪl] n. 纺织品，织物，纺织业
drug [drʌg] n. 药，毒品
plastic ['plæstɪk] n. 塑料，塑料学；adj. 塑料的，可塑的
forensic [fə'renzɪk] adj. 法医的，法院的，公开辩论的
clinical ['klɪnɪkl] adj. 临床的
vital ['vaɪt(ə)l] adj. 必要的，至关重要的
material [mə'tɪərɪəl] n. 布料，原料，材料
functional ['fʌŋkʃənl] adj. 功能的，有多种用途的
definition [ˌdefɪ'nɪʃn] n. 定义，释义
indicate ['ɪndɪkeɪt] v. 表明，暗示，指示，象征
identification [aɪˌdentɪfɪ'keɪʃn] n. 识别，身份证明
species ['spiːʃiːz] n. 物种，种类
determination [dɪˌtɜːmɪ'neɪʃ(ə)n] n. 决心，决定，查明
particular [pə'tɪkjələ(r)] adj. 特定的，特殊的；n. 细节
nickel ['nɪk(ə)l] n. 镍
steel [stiːl] n. 钢，钢铁
rely [rɪ'laɪ] v. 依靠，信赖
require [rɪ'kwaɪə(r)] v. 需要，要求，规定
desire [dɪ'zaɪə(r)] v. 渴望，被某人吸引
approach [ə'prəʊtʃ] n. 方法，方式，接近，来临，途径，道路
analyst ['ænəlɪst] n. 分析者，化验员
accurate ['ækjərət] adj. 正确的，精确的

precise [prɪˈsaɪs] *adj*. 清晰的，正规的，精密的
instrument [ˈɪnstrəmənt] *n*. 器械，乐器，仪器，工具
destroy [dɪˈstrɔɪ] *v*. 毁坏，毁掉某人，杀死
colleague [ˈkɒliːɡ] *n*. 同事，同行
metallurgy [məˈtælədʒɪ] *n*. 冶金，冶金学，冶金术
biology [baɪˈɒlədʒɪ] *n*. 生物学
geology [dʒiˈɒlədʒɪ] *n*. 地质学
limitation [ˌlɪmɪˈteɪʃn] *n*. 限制，局限，极限
provide [prəˈvaɪd] *v*. 提供，规定

Expressions and Technical Terms

analytical chemistry 分析化学
qualitative characterization 定性特征
quantitative characterization 定量特征
qualitative analysis 定性分析
quantitative analysis 定量分析
chemical reaction 化学反应
classical method 经典方法
wet chemical method 湿化学法
analytical field 分析领域
volumetric analysis 容量分析法
gravimetric analysis 重量分析法
material science 材料科学
food science 食品科学

Exercises

A. Translate the following into English.

1. 准确的   2. 精密的   3. 分析化学   4. 化学反应

5. 定性分析   6. 定量分析   7. 材料科学   8. 食品科学

B. Decide whether the following statements are true (T) or false (F). Write T for true and F for false in each blank.

(   ) 1. Almost every day your life is probably affected in some way by chemical analysis.

(   ) 2. The identification of the chemical compounds or elements present in a sample that tells us what is present is called quantitative analysis.

(   ) 3. The determination of how much of a particular compound or element is present in a sample, or the determination of how much of a species is present in a sample is called qualitative analysis.

(   ) 4. Many of the volumetric methods have been transferred to automated instruments.

(　　) 5. It may be necessary to analyze samples without destroying them.

C. Fill in the blanks.

1. The most functional definition of analytical chemistry is that it is "the _____ and _____ characterization of matter".

2. Qualitative analysis is the identification of one or more chemical species _____ in a material.

3. Quantitative analysis is the determination of the exact _____ of a chemical species present in a sample.

4. "Wet chemical methods" usually required that a part of the _____ is taken, _____ in a suitable solvent if necessary and the desired _____ is carried out. The most important analytical fields based on this approach were _____ and _____ analysis.

5. Classical analysis and instrumental analysis are similar in many respects, such as in the need for proper _____, sample _____, assessment of _____ and _____, and proper _____ keeping.

D. Translate following sentences into Chinese.

1. These examples include techniques for monitoring pollutants in air or water and methods for detecting bacteria or contaminants in our food, textiles, drugs, plastics, and metals.

2. The field of chemistry that deals with the use and development of tools and processes for examining and studying chemical substances is known as analytical chemistry.

3. Volumetric and gravimetric analysis require a high degree of skill and attention to details on the part of the analyst if accurate and precise results are to be obtained.

4. The analytical chemist must not only know and understand analytical chemistry and instrumentation, but must also be able to serve as a problem solver to colleagues in other scientific areas.

Lesson 20
The Function of Analytical Chemistry
分析化学的应用

Warming up

分析化学是研究物质的组成、含量、结构和形态等化学信息的分析方法及理论的一门科学，是化学学科的一个重要分支；是鉴定物质中含有哪些组分，即物质由什么组分组成，测定各种组分的相对含量，研究物质的分子结构的科学。

分析化学不仅对化学各学科的发展起着重要作用，而且在许多领域中都有广泛的应用（都需要分析化学的理论、知识和技术）。

1. 化学学科： 只要涉及物质及其变化的研究都需要使用分析化学的方法，如：质量不灭定律的证实（18 世纪中叶）、原子量的测定（19 世纪前半期）、门捷列夫周期律的创建（19 世纪后半期）、有机合成、催化机理和溶液理论等的确证。

2. 医药卫生： 临床医学中用于诊断和治疗的临床检验；预防医学中环境检测、职业中毒检验、营养成分分析等；法医学的法医检验、药学领域的药物成分含量的测定、药物药代动力学及新药的药物分析等；水中三氮（NH_3、HNO_2、HNO_3）的测定；水中有毒物质的测定（Pb、Hg、HCN 等）；食品、蔬菜等中维生素 C 的测定，农药残留量的检测；血液中有毒物质的测定；血液中药物浓度的分析；血液、头发中微量元素的分析；等等。

3. 生命科学： 确定糖类、蛋白质、脱氧核糖核酸（DNA）、酶以及各种抗原抗体、激素及激素受体的组成、结构、生物活性及免疫功能等，常用分光光度法、化学发光法、色谱法等。

4. 工业： 资源勘探，生产原料、中间体、产品的检验分析，工艺流程的控制，产品质量的检验，"三废"的处理等。

5. 农业： 水土成分调查，农产品质量检验，细胞工程、基因工程、发酵工程等。

6. 国防： 核武器的燃料、武器结构材料、航天材料的研究等。

Text

Chemical analysis is an indispensable servant of modern technology whilst it partly depends on that modern technology for its operation. In modern technology, it is impossible to overestimate the importance of chemical analysis. Some of the major areas of its application are listed below.

Fundamental Research

The first step in unravelling the details of an unknown system frequently involves the identification of its constituents by qualitative chemical analysis. Following up investigations usually require structural information and quantitative measurements. This pattern appears in such diverse areas as the formulation of new drugs, the examination of meteorites, and study on the results of heavy ion bombardment by nuclear physicists.

Product Development

The design and development of a new product will often depend upon establishing a link between its chemical composition and its physical properties or performance. Typical examples are the development of alloys and of polymer composites.

Product Quality Control

Most manufacturing industries require a uniform product quality. To ensure that this requirement is met, both raw materials and finished products are subjected to extensive chemical analysis. On the one hand, the necessary constituents must be kept at the optimum levels, while on the other hand impurities such as poisons in foodstuffs must be kept below the maximum allowed by law.

Monitoring and Control of Pollutants

Residual heavy metals and organo-chlorine pesticides represent two well known pollution problems. Sensitive and accurate analysis is required to enable the distribution and level of a pollutant in the environment to be assessed, and routine chemical analysis is important in the control of industrial effluents.

New Words

function [ˈfʌŋkʃn] *n.* 功能
indispensable [ˌɪndɪˈspensəbl] *adj.* 不可缺少的，绝对必要的
servant [ˈsɜːvənt] *n.* 仆人，雇工
whilst [waɪlst] *conj.* 与……同时，然而，尽管
operation [ˌɒpəˈreɪʃ(ə)n] *n.* 活动，使用，运算，运行
overestimate [ˌəʊvərˈestɪmeɪt] *vt.* 对……作过高的评价，对……估计过高
application [ˌæplɪˈkeɪʃn] *n.* 申请，请求，运用
unravel [ʌnˈrævl] *vt.* 解开，阐明
investigation [ɪnˌvestɪˈɡeɪʃn] *n.* 调查，学术研究
pattern [ˈpæt(ə)n] *n.* 模式，范例，图案，模型，样品
diverse [daɪˈvɜːs] *adj.* 形形色色的，不同的
formulation [ˌfɔːmjʊˈleɪʃn] *n.* 配方，规划，公式化
meteorite [ˈmiːtɪəraɪt] *n.* 陨星，陨石
establish [ɪˈstæblɪʃ] *v.* 创建，证实
alloy [ˈælɔɪ] *n.* 合金
composite [ˈkɒmpəzɪt] *n.* 合成物，混合物，复合材料

extensive [ɪkˈstensɪv] *adj.* 广阔的，广泛的，大量的
maximum [ˈmæksɪməm] *n.* 最大量
monitor [ˈmɒnɪtə(r)] *v.* 监控，监听
pollutant [pəˈluːtənt] *n.* 污染物
residual [rɪˈzɪdjʊəl] *adj.* 残余的，残留的
distribution [ˌdɪstrɪˈbjuːʃn] *n.* 分配，分布
environment [ɪnˈvaɪrənmənt] *n.* 生活环境，自然环境
assess [əˈses] *v.* 评估，估价，估算
routine [ruːˈtiːn] *adj.* 常规的，日常的

Expressions and Technical Terms

analytical chemistry 分析化学
chemical analysis 化学分析
modern technology 现代技术
fundamental research 基础研究
qualitative chemical analysis 化学定性分析
structural information 结构信息
quantitative measurement 定量测量
heavy ion bombardment 重离子轰击
nuclear physicist 核物理学家
product development 产品研发
product quality control 产品质量控制
the optimum level 最佳水平
heavy metal 重金属
organo-chlorine pesticide 有机氯类农药
sensitive and accurate analysis 灵敏和准确的分析
industrial effluent 工业废水

Exercises

A. Translate the following into English.
1. 基础研究   2. 产品研发   3. 产品质量控制   4. 现代技术

5. 污染物监测和控制 6. 最佳水平   7. 重金属   8. 工业废水

B. Decide whether the following statements are true（T）or false（F）. Write T for true and F for false in each blank.

（   ）1. In modern technology，it is impossible to overestimate the importance of analysis.

（   ）2. The second step in unravelling the details of an unknown system frequently involves the identification of its constituents by qualitative chemical analysis.

（   ）3. Typical examples of product development are the development of alloys and of polymer composites.

(　　) 4. To ensure that uniform product quality requirement is met, both raw materials and finished products are subjected to extensive chemical analysis.

(　　) 5. Sensitive and accurate analysis is required to enable the distribution and level of a pollutant in the environment to be assessed.

C. Comprehension exercises.

1. How many functions does the analytical chemistry have?

2. Is the analytical chemistry often used in monitoring and controlling pollutants? Please give some examples about it.

Reading Material 阅读材料

Chemical Manufacturing Process

Typical chemical processes are shown in following figure.

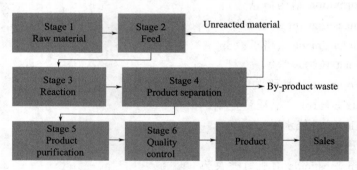

Stage 1. Raw material

The raw materials are often transported by ship, by road and rail. They are made to hold a few hours or a few days storage.

Stage 2. Feed

Some raw materials must be pure or in the right form before they are fed to the reactor. For example, some liquid feeds will need to be vaporized. Solids may need crushing, grinding and screening.

Stage 3. Reaction

The reaction stage is the heart of a chemical manufacturing process. In the reactor, the products will be formed and some by-products and unwanted compounds will also be formed.

Stage 4. Product separation

In this stage, the wastes and by-products are separated out. The unreacted material will be recycled to the reactor.

Stage 5. Product purification

Before sale, the main product will usually need purification. Some by-products may also be purified for sale.

Stage 6. Quality control

Analysis and inspection technology is the 'eyes' of industrial production, because it can check product quality up to standard, and guide production process adjustment and innovation.

New Words

manufacture [ˌmænjʊˈfæktʃə(r)] vt. 制造，加工
figure [ˈfɪɡə(r)] n. 图形
stage [steɪdʒ] n. 阶段，时期
transport [ˈtrænspɔːt] vt. 传送，运输
storage [ˈstɔːrɪdʒ] n. 储藏，存储
feed [fiːd] v. 进料，加料
separation [ˌsepəˈreɪʃn] n. 分离，分开
reactor [rɪˈæktə(r)] n. 反应器
vaporize [ˈveɪpəraɪz] v. （使）蒸发
crush [krʌʃ] vt. 压碎，碾碎
grind [ɡraɪnd] v. 磨碎，碾碎
screen [skriːn] v. 筛分，筛选
purification [ˌpjʊərɪfɪˈkeɪʃən] n. 净化，纯化
purify [ˈpjʊərɪfaɪ] v. 净化，纯化

Expressions and Technical Terms

chemical manufacturing process 化学生产过程
raw material 原材料
be fed to 被加入……
by-product 副产品
be separated out 被分离出……
be recycled to 被循环到……
quality control 质量控制

Exercises

A. Translate the following into English.

1. 化学生产过程　　2. 原材料　　3. 产品　　4. 副产品

5. 产品分离　　6. 纯化　　7. 未反应的物质　　8. 粉碎

9. 研磨　　10. 过筛　　11. 进料　　12. 质量控制

B. Fill in the blanks.

1. The raw materials are made to hold a few hours or a few days _____ .

2. Some solid raw materials may need _____, _____ and _____.

3. In the reactor, _____ will be formed and some _____ and _____ will also be formed.

4. In the product separation, the unreacted material will be _____ to the reactor.

5. Before sale, the main product will usually need _____.

6. Analysis and inspection technology is the '_____' of industrial production.

Unit 1
Chemical Analysis
化学分析

Lesson 21
Titration 滴定法

Warming up

滴定分析法（或称容量分析法）：是化学分析法的一种，将一种已知其准确浓度的试剂溶液（称为标准溶液）滴加到被测物质的溶液中（或者是将被测物质的溶液滴加到标准溶液中），直到所加的试剂与被测物质按化学计量关系定量反应完全时为止，然后根据所用试剂溶液的浓度和体积可以求得被测组分的含量。

滴定分析法是一种简便、快速和应用广泛的定量分析方法，在常量分析中有较高的准确度。

标准滴定溶液：准确滴加到被测溶液中的已知准确浓度的试剂溶液（标准溶液），在滴定分析中，称为滴定剂。

基准物质：能直接配成标准溶液或标定溶液浓度的物质。基准物质须具备的条件：（1）组成恒定，实际组成与化学式符合；（2）纯度高，一般纯度应在 99.5% 以上；（3）性质稳定，保存或称量过程中不分解、不吸湿、不风化、不易被氧化等；（4）具有较大的摩尔质量，称取量大，称量误差小；（5）使用条件下易溶于水（或稀酸、稀碱）。

滴定：滴定分析时将标准溶液通过滴定管逐滴加到锥形瓶中进行测定的过程。滴定分析，以及滴定分析法即因此而得名。

化学计量点：当滴加滴定剂的量与被测物质的量之间，正好符合化学反应式所表示的化学计量关系时，即滴定反应达到化学计量点，简称计量点，又称等当点。

指示剂：指示化学计量点到达而能改变颜色的一种辅助试剂。

滴定终点：在化学计量点时，没有任何外部特征，而必须借助于指示剂变色来确定停止滴定的点。即把这个指示剂变色点称为滴定终点，简称终点。

终点误差：滴定终点与等当点往往不一致，由此产生的分析误差，称为终点误差。

适合滴定分析的化学反应，应该具备以下几个条件：

（1）反应必须按方程式定量地完成，通常要求在 99.9% 以上，这是定量计算的基础。

（2）反应能够迅速地完成（有时可加热或用催化剂以加速反应）。

（3）共存物质不干扰主要反应，或用适当的方法消除其干扰。

（4）有比较简便的方法确定计量点（指示滴定终点）。

Text

Titration is one of the most useful and accurate analytical techniques. It is fairly rapid. In a titration, the test substance (analyte) reacts with a solution of known concentration. This is called a standard solution, and it is generally added from a buret. The added solution is called titrant. The volume of titrant, which just completely reacts with analyte is measured. Since the concentration, the volume and the reaction are all known, the amount of analyte can be calculated.

There are many requirements in a titration.

(1) There is a known reaction between the analyte and the titrant. For example, $CH_3COOH + NaOH \longrightarrow CH_3COONa + H_2O$.

(2) The reaction should be rapid. Most ionic reactions are very rapid.

(3) There are no side reactions. If there are interfering substances, these must be removed. In the above example, there should be no other acids.

(4) When the reaction is complete, there is a marked change in some properties of the solution.

New Words

titration [taɪˈtreɪʃn] *n.* 滴定
accurate [ˈækjərət] *adj.* 正确的,精确的
analyte [ænəˈlaɪt] *n.* (被)分析物
buret [bjʊəˈret] *n.* 滴定管,玻璃量管
titrant [ˈtaɪtrənt] *n.* 滴定剂
calculate [ˈkælkjʊleɪt] *v.* 计算

Expressions and Technical Terms

analytical technique 分析技术
known concentration 已知浓度
standard solution 标准溶液
ionic reaction 离子反应
side reaction 副反应
interfering substance 干扰物质

Exercises

A. Translate the following into English.

1. 滴定　　　　2. 滴定剂　　　　3. 副反应　　　　4. 离子反应

5. 标准溶液　　6. 已知浓度　　　7. 干扰物质　　　8. 分析物

B. Fill in the blanks.

1. Titration is one of the most useful and accurate _____ .

2. A standard solution is generally added from a _____.

3. In a titration, there should be a _____ between the analyte and the titrant.

4. In a titration, the reaction should be _____.

5. There are no side reactions. If there are interfering substances, these must be _____.

C. Choose the best answer to each question.

1. In a titration, the test substance (　　) reacts with a solution of known concentration. This is called a (　　).

　(A) analyte　　　(B) standard solution　　　(C) indicator

2. Since the concentration is known and since the volume is known and since the reaction is known, the (　　) of analyte can be calculated.

　(A) amount　　　(B) volume　　　(C) weight

3. If there are interfering substances, these must be (　　).

　(A) removed　　　(B) oxidized　　　(C) filtered

4. When the reaction is complete, there is a (　　) in some properties of the solution.

　(A) marked change　　(B) disappearance　　(C) appearance

Lesson 22
The Types of Titration
滴定法的类型

Warming up

滴定分析法：又叫容量分析法，将已知准确浓度的标准溶液，滴加到被测溶液中（或者将被测溶液滴加到标准溶液中），直到所加的标准溶液与被测物质按化学计量关系定量反应为止，然后测量标准溶液消耗的体积，根据标准溶液的浓度和所消耗的体积，算出待测物质的含量。这种定量分析的方法是一种简便、快速和应用广泛的定量分析方法，在常量分析中有较高的准确度。滴定分析是建立在滴定反应基础上的定量分析法。

根据标准溶液和待测组分间的反应类型的不同，分为如下四类：

1. 酸碱滴定法：以质子传递反应为基础的一种滴定分析方法。例如氢氧化钠测定乙酸。

2. 配位滴定法：以配位反应为基础的一种滴定分析方法。例如 EDTA 测定水的硬度。

3. 氧化还原滴定法：以氧化还原反应为基础的一种滴定分析方法。例如高锰酸钾测定铁含量。

4. 沉淀滴定法：以沉淀反应为基础的一种滴定分析方法。例如食盐中氯的测定。

Text

There are four types of titration methods.

（1）Acid-base titrations

Many compounds are either acids or bases. They can be titrated with a strong base or a strong acid. The end points of these titrations are easy to detect by means of an indicator.

（2）Precipitation titrations

The titrant forms an insoluble product with the analyte. For example, we can use silver nitrate solution to titrate with chlorine ion. Again, indicators can be used to detect the end points.

（3）Complexometric titrations

In complexometric titrations, the titrant is complexing agent. It can form a water soluble complex with the analyte, a metal ion. EDTA is one of the most useful complexing agents. In the titration, it can react with many metal ions.

（4）Oxidation-reduction titrations

In these titrations, an oxidizing agent reacts with a reducing agent, or vice versa. An oxidizing agent gains electrons and a reducing agent loses electrons in a reaction between them. Indicator can be used to detect the end point.

New Words

titrate [taɪˈtreɪt] v. 滴定，用滴定法测量
complexometric [ˌkɒmpleksˈɒmɪtrɪk] adj. 配位的
complexing [kɑːmˈpleksɪŋ] n. 配位
electron [ɪˈlektrɒn] n. 电子

Expressions and Technical Terms

acid-base titration 酸碱滴定
strong acid 强酸
strong base 强碱
end point 终点
by means of 依靠，借助于
precipitation titration 沉淀滴定
complexometric titration 配位滴定
silver nitrate 硝酸银
complexing agent 配位剂
oxidation-reduction titration 氧化还原滴定
oxidizing agent 氧化剂
reducing agent 还原剂
vice versa 反之亦然

Exercises

A. Translate the following into English.

1. 酸碱滴定　　2. 配位滴定　　3. 沉淀滴定　　4. 氧化还原滴定

5. 终点　　6. 氧化剂　　7. 还原剂　　8. 配位剂

9. 强酸　　10. 强碱

B. Fill in the blanks.

1. The end points of acid-base titrations are easy to detect by means of an _____ .

2. In precipitation titrations, the titrant forms an _____ product with the analyte.

3. EDTA can form a water soluble complex with the analyte, a metal ion. It is one of the most useful _____ .

4. An oxidizing agent _____ electrons and a reducing agent _____ electrons in a

reaction between them.

5. There are four types of titration methods, _____ titrations, _____ titrations, _____ titrations and oxidation-reduction _____ .

C. Answer the following questions.
1. How many types of titration methods are there?
2. What can be used to detect the end points in an acid-base titration?
3. What does an oxidizing agent react with in a reduction-oxidation titration?
4. What is the titrant in complexometric titrations?

Lesson 23
Use of Burettes 滴定管的使用

Warming up

滴定管是滴定分析法所用的主要量器，可分为两种：酸式滴定管、碱式滴定管。

校准： 滴定管的容积与其所标出的体积并非完全一致，在准确度要求较高的分析工作中须进行校准。由于玻璃具有热胀冷缩的特性，在不同温度下，滴定管的体积不同。校准时，必须规定一个共同的温度值，这一规定温度值为标准温度。国际上规定玻璃容量器皿的标准温度为 20℃，即在校准时都将玻璃容量器皿的容积校准到 20℃ 时的实际容积。

滴定管的校准与容量瓶的校准一样，都采用称量水法，即根据纯水在不同温度下具有不同的密度，称量测量温度下滴定管不同刻度处水的质量，根据 $V=m/\rho$，计算该温度下纯水的体积，即为该滴定管在该刻度处的真实容积。

使用注意事项

1. 两检：一是检查滴定管是否破损；二是检查滴定管是否漏水，如是酸式滴定管还要检查玻璃塞旋转是否灵活。

2. 三洗：滴定管在使用前必须洗净。

一洗：当没有明显污染时，可以直接用自来水冲洗。如果其内壁沾有油脂性污物，则可用肥皂液、合成洗涤液或碳酸钠溶液润洗，必要时把洗涤液先加热，并浸泡一段时间。所有洗涤剂在洗涤容器后，都要倒回原来盛装的瓶中。无论用肥皂液、洗液等都需要用自来水充分洗涤。

二洗：用蒸馏水淌洗 2~3 次，每次用 5~10mL 蒸馏水。

三洗：用欲装入的标准溶液最后淌洗 2~3 次，每次用 5~10mL 溶液，以除去残留的蒸馏水，保证装入的标准溶液与试剂瓶中的溶液浓度一致。

3. 标准溶液的装入。

4. 排气：即排除滴定管下端的气泡。

5. 滴定管的读数：手拿滴定管上端无溶液处使滴定管自然下垂，并将滴定管下端悬挂的液滴除去后，眼睛与液面在同一水平面上，进行读数，要求读准至小数点后两位。普通滴定管装无色溶液或浅色溶液时，读取弯月面下缘最低点处；溶液颜色太深时，无法观察下缘时，可读液面两侧的最高点。读数卡是用涂有黑色的长方形（约 3cm×1.5cm）的白纸制成的。读数卡放在滴定管背后，使黑色部分在弯月面下约 1mm 处，即可看到弯月面的反射层成为黑色，然后读此黑色弯月面下缘的最低点。溶液颜色深而读取最上缘时，就可以用白纸作为读数卡。在装好标准溶液或放出标准溶液后，都必须等 1~2 分钟，使溶液完全从器壁上流下后再读数。

6. 滴定操作要求：规范。

Text

It is necessary for the student to distinguish carefully between the volume that a vessel holds and the volume that it delivers. If exactly one liter of liquid is poured into a clean dry flask, the amount of liquid which can be poured from the flask will be less than one liter. An appreciable quantity of the liquid will cling to the sides of the flask. The quantity of liquid delivered by the flask is held, as well as other factors. If it is desired that the flask delivers exactly one liter, the neck of the flask must be marked at points. They will vary depending upon the characteristics of the fluid in the flask. The neck, however, may be marked at a point which shows that the flask contains exactly one liter irrespective of the liquid in the flask. Volumetric flasks are marked in this way to show the amount which is contained by the flask, not the amount which the flask will deliver. Burettes and transfer (volumetric) pipettes, on the other hand, are marked so that they will deliver a definite volume of liquid.

Usually two 50mL burettes are furnished each student in quantitative analysis. One of these will be a Geissler burette with a glass or teflon stopcock, and the other will be a Mohr burette which is fitted with a glass bead in rubber tubing to control the flow of liquid. The Mohr burette is used for basic titrants such as sodium hydroxide and the Geissler burette is used for practically all other titrants.

During titrations, the burette stopcock is normally operated with the left hand. This leaves the right hand free for stirring the liquid or rotating the titration flask. Left-handed students should reverse this and use the right hand for operating the stopcock. Proper control of the stopcock to ensure delivery of small or large volumes accurately requires relaxed muscular control and extensive practice. Muscular tension often causes the student to push the stopcock out of the burette and thus ruin the titration.

There are many errors which can be involved in the use of burette, especially by the beginning student in quantitative analysis:

(1) Dirty burettes can cause an error because the volume of a liquid which clings to a dirty glass surface is different from the volume which clings to a clean surface.

(2) If the liquid is allowed to flow out of the burette rapidly, the student should wait at least 1 min. An error of as much as 0.20mL can result from reading the position of the meniscus too soon.

(3) In filling the burette, care must be taken to ensure that no bubble of air remains in the tip.

(4) The burette should either be dried before it is filled with the solution, or else it should be rinsed at least three times with small volumes of the solution before filling.

(5) The student should position his eyes so that the back of the milliliter mark above the meniscus appears to be below the front portion. The back of the milliliter mark below the meniscus appears to be above the front portion.

(6) A piece of white paper or a piece of white paper upon which a large black mark (with smooth upper edge) is made may be placed behind the burette. It can sharpen the bottom of the

meniscus to aid in the estimation of the correct reading.

（7）It is often necessary when approaching the end point of a titration to add a partial drop of titrant. To do this allows the partial drop to form on the tip of the burette and then to touch the tip to the side of the titration flask. The flask can then be tilted so that the added amount is mixed with the solution or it can be washed down with water from the wash bottle. The burette tip should revere to be rinsed since this may dilute the titrant in the tip and cause an error.

New Words

burette [bjʊˈret] n. 滴定管
vessel [ˈvesl] n. 容器，船，飞船
deliver [dɪˈlɪvə(r)] v. 发表，宣布，递送，交付，运载
liter [ˈliːtə(r)] n. 升（体积单位）
pour [pɔː(r)] v. 倾倒，倒
appreciable [əˈpriːʃəb(ə)l] adj. 可预见的，可估计的，可察觉的
cling [klɪŋ] v. 抓紧，粘住，依附
irrespective [ˌɪrɪˈspektɪv] adj. 不考虑的，不顾的，无关的
pipette [pɪˈpet] n. 吸液管
furnish [ˈfɜːnɪʃ] vt. 布置，提供，装修
teflon [ˈteflɒn] n. 聚四氟乙烯
stopcock [ˈstɒpkɒk] n. 活塞，活栓，旋塞阀
bead [biːd] n. 珠子
muscular [ˈmʌskjələ(r)] adj. 肌肉的，健壮的
ruin [ˈruːɪn] v. 毁坏，破坏
error [ˈerə(r)] n. 错误，过失
meniscus [məˈnɪskəs] n. 新月，半月板
tip [tɪp] n. 尖/顶端
rinse [rɪns] vt. 漂洗，冲洗
portion [ˈpɔːʃ(ə)n] n. 一部分
edge [edʒ] n. 边线
estimation [ˌestɪˈmeɪʃn] n. 评价，判断，估算
approach [əˈprəʊtʃ] v. 接近，临近
partial [ˈpɑːʃ(ə)l] adj. 部分的
tilt [tɪlt] vt. 使倾斜
revere [rɪˈvɪə(r)] vt. 崇敬，尊崇，敬畏

Expressions and Technical Terms

depend upon… 依靠，取决于……
volumetric flask 容量瓶
transfer（volumetric）pipette 移液管
quantitative analysis 定量分析

Geissler burette 盖斯勒滴定管
Mohr burette 莫尔滴定管
rubber tubing 橡胶管
muscular control 肌肉控制
extensive practice 全面的练习
at least 至少
wash bottle 洗瓶

Exercises

A. Translate the following into English.

1. 滴定管　　　2. 容量瓶　　　3. 移液管　　　4. 洗瓶

5. 滴定终点　　6. 取决于……　7. 误差　　　　8. 全面的练习

B. Choose the best answer to each question.

1. (　　) burettes can cause an error.
(A) Dirty　　　(B) Big　　　(C) Long

2. If the liquid is allowed to flow out of the burette rapidly, the student should wait at least (　　) min.
(A) 0.5　　　(B) 1　　　(C) 2

3. In filling the burette, care must be taken to ensure that no (　　) remains in the tip.
(A) bubble of air　　(B) titrant　　(C) bead

4. The burette may be rinsed at least (　　) times with small volumes of the solution before filling.
(A) 2　　　(B) 3　　　(C) 4

5. A piece of (　　) paper may be placed behind the burette to sharpen the bottom of the meniscus to aid in the estimation of the correct reading.
(A) white　　(B) black　　(C) grey

C. Decide whether the following statements are true (T) or false (F) according to the text. Write T for true and F for false in each blank.

(　　) 1. If exactly one liter of liquid is poured into a clean dry flask, the amount of liquid which can be poured from the flask will be equal to one liter.

(　　) 2. Volumetric flasks are marked in this way to show the amount which is contained by the flask, not the amount which the flask will deliver.

(　　) 3. Burettes and transfer (volumetric) pipettes are marked so that they will deliver a definite volume of liquid.

(　　) 4. Usually two 25mL burettes are furnished each student in quantitative analysis.

(　　) 5. During titrations, the burette stopcock is normally operated with the right hand. This leaves the left hand free for stirring the liquid or rotating the titration flask.

Lesson 24
Acid-Base Titrations 酸碱滴定法

 Warming up

酸碱滴定法： 是指利用酸和碱在水中以质子转移反应为基础的滴定分析方法。可用于测定酸、碱和两性物质。其基本反应为 $H^+ + OH^- \rightleftharpoons H_2O$，也称中和法，是一种利用酸碱反应进行定量分析的方法。用酸作滴定剂可以测定碱，用碱作滴定剂可以测定酸。

它是一种用途极为广泛的分析方法。酸碱滴定法在工农业生产和医药卫生等方面都有非常重要的意义。"三酸二碱"是重要的化工原料，它们都用此法分析。

有的非酸或非碱物质经过适当处理可以转化为酸或碱，然后也可以用酸碱滴定法测定。例如，测定有机物的含氮量。

滴定曲线： 指在容量分析中，以滴定过程中所用标准溶液的体积对溶液的某些特性的相应改变做成的曲线。酸碱滴定曲线特性为 pH 值。

酸碱指示剂： 用于酸碱滴定的指示剂。是一类结构较复杂的有机弱酸或有机弱碱，它们在溶液中能部分电离成指示剂的离子和氢离子（或氢氧根离子），并且由于结构上的变化，它们的分子和离子具有不同的颜色，因而在 pH 不同的溶液中呈现不同的颜色。

酸碱滴定需要选择合适的指示剂来指示滴定终点，变色范围全部或者部分落在滴定突跃范围内的指示剂都可以用来指示终点。常见的酸碱指示剂有甲基橙、甲基红、溴酚蓝、溴甲酚蓝、酚酞、百里酚酞等。

Text

A titration, or "titrimetric analysis", is a procedure in which the quantity of an analyte in a sample is determined by adding a known quantity of a reagent that reacts completely with the analyte in a well-defined manner. The reagent that is combined with the analyte in this method is known as the titrant. This approach involves the use of a buret to carefully deliver a known volume of titrant to a sample. During this process, the titrant/sample mixture is examined for a change in color, pH, or other measurable properties. It can be used to signal when the analyte has been completely consumed by the titrant.

An acid-base titration is a special type of titration in which the reaction of an acid with a base is used for measuring an analyte. For example, if the analyte is an acid such as hydrochloric acid, the titrant would be a base like sodium hydroxide. Similarly, if the analyte is a base such as sodium hydroxide, the titrant would be an acid like hydrochloric acid.

A plot of the measured response versus the amount of added titrant during a titration is called a titration curve. The point in this curve at which exactly enough titrant has been added to react with all of the analyte is known as equivalent point. Once we have determined the amount of titrant that must be added to reach the equivalent point, we can determine the amount of analyte. It was present in the original sample by using the known stoichiometry for the reaction of the titrant with the analyte. Besides having a well-characterized reaction between the analyte and titrant, it is also desirable in a titration for this reaction to be fast and to have a large equilibrium constant. These properties help to ensure that the titrant will quickly and completely combine with the analyte as the titrant is added to the sample.

Titrations are usually inexpensive to conduct and require only simple, standard laboratory equipment. Titrations are also capable of providing both good accuracy and precision during a chemical analysis. Acid-base reactions are especially well suited for titration, because these reactions tend to have large equilibrium constants and fast reaction rates.

New Words

titrate [taɪˈtreɪt] v. 滴定，用滴定法测量
titration [taɪˈtreɪʃn] n. 滴定
quantity [ˈkwɒntəti] n. 数量，大量，众多
analyte [ˈænəlaɪt] n. （被）分析物，分解物
manner [ˈmænə(r)] n. 方式，做法，规矩
titrant [ˈtaɪtrənt] n. 滴定剂，滴定（用）标准液
signal [ˈsɪɡnəl] v. 发信号，示意
plot [plɒt] n. 情节，图表；v. 计划，标绘出位置，绘制
response [rɪˈspɒns] n. 答复，反应
versus [ˈvɜːsəs] prep. （表示两队或双方对阵）对，（比较两种不同想法、选择等）与……相对，对抗
equivalence [ɪˈkwɪvələns] n. 相等，对等
stoichiometry [ˌstɔɪkɪˈɒmɪtri] n. 化学计算（法），化学计量学
desirable [dɪˈzaɪərəb(ə)l] adj. 令人满意的，值得拥有的，可取的
equilibrium [ˌiːkwɪˈlɪbriəm] n. 平衡，均势，平静
accuracy [ˈækjərəsi] n. 精确（性），准确（性），准确无误
precision [prɪˈsɪʒn] n. 精确度，准确（性）
especially [ɪˈspeʃəli] adv. 尤其，特别

Expressions and Technical Terms

acid-base titration 酸碱滴定
titrimetric analysis 滴定分析（法）
a known quantity of... 已知一定量的……
hydrochloric acid 盐酸

sodium hydroxide 氢氧化钠
titration curve 滴定曲线
equivalent point 等当点
equilibrium constant 平衡常数
laboratory equipment 实验室设备

Exercises

A. Translate the following into English.

1. 酸碱滴定　　　2. 滴定分析法　　　3. 一定量的……　　　4. 滴定曲线

5. 平衡常数　　　6. 实验室设备　　　7. 化学计量点　　　8. 一定体积的……

B. Cloze.

A _____, or "titrimetric analysis", is a procedure in which the _____ of an analyte in a sample is determined by adding a _____ quantity of a reagent that reacts _____ with the analyte in a well-defined _____. The reagent that is combined with the analyte in this method is known as the _____. This approach involves the use of a _____ to carefully deliver a known _____ of titrant to a sample. During this process, the titrant/sample mixture is examined for a _____ in color, pH, or other measurable properties that can be used to _____ when the analyte has been completely consumed by the titrant.

C. Translate following sentences into Chinese.

1. A plot of the measured response *versus* the amount of added titrant during a titration is called a titration curve.

2. The point in this curve at which exactly enough titrant has been added to react with all of the analyte is known as equivalent point.

3. Acid-base reactions are especially well suited for titration, because these reactions tend to have large equilibrium constants and fast reaction rates.

D. Answer the following questions.

1. What are the advantages of an acid-base titration?
2. What kind of reaction is suitable to titration?

Lesson 25
Complexometric Titration: EDTA Titration Procedure
络合滴定法：EDTA 络合滴定的过程

Warming up

络合滴定法是以络合反应为基础的滴定分析方法。

在络合反应中，提供配位原子的物质称为配位体。

乙二胺四乙酸及其二钠盐简称 EDTA，是含有羧基和氨基的螯合剂，能与许多金属离子形成稳定的螯合物。在化学分析中，它除了用于络合滴定以外，在各种分离、测定方法中，还广泛地用作掩蔽剂。

滴定方式

1. 直接滴定法：这是络合滴定中最基本的方法。这种方法是将被测物质处理成溶液后，调节酸度，加入指示剂（有时还需要加入适当的辅助络合剂及掩蔽剂），直接用 EDTA 标准溶液进行滴定，然后根据消耗的 EDTA 标准溶液的体积，计算试样中被测组分的含量。

2. 返滴定法：是将被测物质制成溶液，调好酸度，加入过量的 EDTA 标准溶液（总量 c_1V_1），再用另一种标准金属离子溶液，返滴定过量的 EDTA（c_2V_2），算出两者的差值，即是与被测离子结合的 EDTA 的量，由此就可以算出被测物质的含量。这种滴定方法，适用于无适当指示剂或与 EDTA 不能迅速络合的金属离子的测定。

3. 置换滴定法：利用置换反应生成等物质的量的金属离子或 EDTA，然后进行滴定的方法，称为置换滴定法。即在一定酸度下，往被测试溶液中加入过量的 EDTA，用金属离子滴定过量的 EDTA，然后再加入另一种络合剂，使其与被测定离子生成一种络合物，这种络合物比被测离子与 EDTA 生成的络合物更稳定，从而把 EDTA 释放（置换）出来，最后再用金属离子标准溶液滴定释放出来的 EDTA。根据金属离子标准溶液的用量和浓度，计算出被测离子的含量。这种方法适用于多种金属离子存在下测定其中一种金属离子。

4. 间接滴定法：有些金属离子（如 Li、Na、K、Rb、Cs、W、Ta 等）和一些非金属离子，由于不能和 EDTA 络合或与 EDTA 生成的络合物不稳定，不便于络合滴定，这时可采用间接滴定的方法进行测定。

Text

If the analyte metal ion forms a stable EDTA complex rapidly, and an end point can be readily detected, a direct titration procedure may be employed. More than thirty metal ions may be so determined. Where the analyte is partially precipitated under the reaction conditions thereby leading to a slow reaction, or where a suitable indicator cannot be found, back titration procedures are used. A measured excess of EDTA is added and the unreacted EDTA is titrated with a standard magnesium or calcium solution. Provided the analyte complex is stronger than the Ca-EDTA or Mg-EDTA complex a satisfactory end point may be obtained with eriochrome black T as indicator. An alternative procedure, where end points are difficult to observe, is to use a displacement reaction. In this case, a measured excess of EDTA is added as its zinc or magnesium complex. Provided the analyte complex is the stronger, the analyte will displace the zinc or magnesium

$$MgY^{2-} + M^{2+} \longrightarrow MY^{2-} + Mg^{2+}$$

The magnesium will be liberated quantitatively and may then be titrated with a standard EDTA solution. Where mixtures of metal ions are analysed, the masking procedures already discussed can be utilized or the pH effect is exploited. A mixture containing bismuth, cadmium and calcium might be analysed by first titrating the bismuth at pH=1-2 followed by the titration of cadmium at an adjusted pH=4 and finally calcium at pH=8. Titrations of this complexity would be most conveniently carried out potentiometrically using the mercury pool electrode.

Most procedures will require an analyte concentration of 10^{-3} mol/dm^3 or more, although with special conditions, notably potentiometric end point detection, the sensitivity may be extended to 10^{-4} mol/dm^3. The analysis mixtures of metal ions necessitate masking and demasking, pH adjustments and selective separation procedures. Areas of application are spread throughout the chemical field from water treatment and the analysis of refined food and petroleum products to the assay of minerals and alloys.

New Words

complexometric [ˌkɒmpleksˈɒmɪtrɪk] *adj.* 络合的
partial [ˈpɑːʃ(ə)l] *adj.* 部分的
alternative [ɔːlˈtɜːnətɪv] *adj.* 替代的，备选的
thereby [ˌðeəˈbaɪ] *adv.* 由此，从而
mask [mɑːsk] *n.* 面具，口罩；*v.* 掩蔽，遮住
exploit [ɪkˈsplɔɪt] *v.* 开发，利用
bismuth [ˈbɪzməθ] *n.* 铋
cadmium [ˈkædmɪəm] *n.* 镉
calcium [ˈkælsɪəm] *n.* 钙
complexity [kəmˈpleksətɪ] *n.* 复杂性
adjust [əˈdʒʌst] *v.* 调整，调节，适应，校准
convenient [kənˈviːnɪənt] *adj.* 实用的，方便的

potentiometric [pəˌtenʃɪəˈmetrɪk] *adj*. 电势测定的，电位计的
electrode [ɪˈlektrəʊd] *n*. 电极，电焊条
notable [ˈnəʊtəbl] *adj*. 值得注意的，显著的，著名的
notably [ˈnəʊtəblɪ] *adv*. 显著地，尤其
necessitate [nəˈsesɪteɪt] *vt*. 使……成为必要，需要，强迫
demask [ˈdiːməsk] *v*. 暴露
refine [rɪˈfaɪn] *vt*. 提炼，改善
assay [əˈseɪ] *n*. 化验，试验
analyte [ˈænəlaɪt] *n*. （被）分析物，分解物

Expressions and Technical Terms

complexometric titration 络合滴定法
direct titration procedure 直接滴定
back titration procedure 返滴定
eriochrome black T 铬黑 T
a measured excess of… 一定量过量的……
the masking procedure 掩蔽过程
masking and demasking 掩蔽和解蔽
pH adjustment 调节 pH 值
lead to… 导致……
petroleum product 石油产品，成品油

Exercises

A. Translate the following into Chinese or English.

1. eriochrome black T   2. separation procedure

3. demasking reagent   4. displacement titration

5. assay of mineral   6. lead to higher result

7. 终点   8. 返滴定   9. 直接滴定   10. 一定量过量的

11. 掩蔽剂   12. 调节 pH 值

B. Decide whether the following statements are true（T）or false（F）according to the text. Write T for true and F for false in each blank.

（　　）1. A direct titration procedure may be used when the metal ions react with EDTA slowly.

（　　）2. The eriochrome black T can be used as the indicator in displacement titration of EDTA provided the analyte complex is stronger than the Ca-EDTA or Mg-EDTA.

(　　) 3. A mixture of bismuth, cadmium and calcium might be analyzed by first titrating the bismuth at pH=1-2 then cadmium at an adjusted pH=4 and finally calcium at pH=8.

(　　) 4. Most procedures of EDTA titration will require an analyte concentration of 10^{-3} mol/dm^3 or less.

(　　) 5. Masking and demasking reagents are necessary to the analysis of mixtures of metal ions in EDTA titrations.

(　　) 6. EDTA titration methods never need separation procedures.

C. Translate the following paragraph into Chinese.

A titrimetric method involves the controlled reaction of a standard reagent in known amounts with a solution of the analyte, in order that the stoichiometric or equivalent point for the reaction between the reagent and the analyte may be located. If the details of the reaction are known and the stoichiometric point is located precisely, the amount of analyte present may be calculated from the known quantity of standard reagent consumed in the reaction.

Lesson 26
Precipitation Titrations
沉淀滴定法

Warming up

沉淀反应： 产生沉淀的化学反应。物质的沉淀和溶解是一个平衡过程，通常用溶度积常数 K_{sp} 来判断难溶盐是沉淀还是溶解。

溶度积常数： 指在一定温度下，在难溶电解质的饱和溶液中，组成沉淀的各离子浓度的乘积为一常数。分析化学中经常利用这一关系，加入相同离子而使沉淀溶解度降低，使残留在溶液中的被测组分小到可以忽略的程度。

沉淀滴定法： 以沉淀反应为基础的一种滴定分析方法。

沉淀滴定法必须满足的条件：①溶解度小，且能定量完成；②反应速率快；③有适当指示剂指示终点；④吸附现象不影响终点观察。虽然可定量进行的沉淀反应很多，但由于缺乏合适的指示剂，因而应用于沉淀滴定的反应并不多，目前比较有实际意义的是银量法。

银量法： 以硝酸银溶液为滴定液，测定能与 Ag^+ 反应生成难溶性沉淀的一种容量分析法。即利用 Ag^+ 与卤素离子的反应来测定 Cl^-、Br^-、I^-、SCN^- 和 Ag^+。银量法共分三种，分别以创立者的姓名来命名：莫尔法、福尔哈德法和法扬斯法。

在经典的定性分析中，几乎一半以上的检出反应是沉淀反应。在定量分析中，它是重量法和沉淀滴定法的基础。沉淀反应也是常用的分离方法，既可将欲测组分分离出来，也可将其他共存的干扰组分沉淀除去。

Text

A precipitation reaction is similar to an acid-base titration and a complexometric titration in which these are often performed as volumetric methods. It uses a measurement of titrant volume to determine the amount of an analyte in a sample. However, these titration methods make use of different types of reactions.

A precipitation titration is a titration method in which the reaction of a titrant with a sample leads to an insoluble precipitate. This method is generally carried out by adding to a sample a known volume of a solution containing a precipitating agent until no further precipitate is formed. For example, a sample containing Ag^+ is titrated with a solution containing a known concentration of Cl^-, leading to formation of insoluble AgCl(s). When the lack of further precipitation or a related signal indicates that the end point has been reached, the delivered volume of titrant (which has a known concentration and reaction stoichiometry with analyte) is used to determine the concentration of the analyte in the original sample.

Most precipitation titrations that are used today are based on either the analysis of silver or the use of Ag^+ as a precipitating agent. The earliest of these was a method developed in 1829, and several improvements were later made in this technique. An example is a titration method that was created in 1874, which is now known as the Volhard method. This technique involves the titration of Ag^+ with thiocyanate (SCN^-) to give solid AgSCN. Other precipitation titrations that can also make use of Ag^+ as a titrant and for end point detection are the Mohr method and Fajans method.

New Words

precipitation [prɪˌsɪpɪˈteɪʃn] n. 沉淀，(雨等) 降落
insoluble [ɪnˈsɒljəb(ə)l] adj. 不能解决的，不溶的
original [əˈrɪdʒənl] adj. 最初的
thiocyanate [ˌθaɪəʊˈsaɪəneɪt] n. 硫氰酸盐

Expressions and Technical Terms

precipitation titration 沉淀滴定
complexometric titration 配位滴定
volumetric method 容量法
precipitating agent 沉淀剂
the Volhard method 福尔哈德法
the Mohr method 莫尔法
the Fajans method 法扬斯法

Exercises

A. Translate the following into Chinese or English.

1. 沉淀滴定法   2. 沉淀剂   3. 容量法   4. thiocyanate

5. Volhard method   6. Mohr method   7. Fajans method

B. Decide whether the following statements are true (T) or false (F) according to the text. Write T for true and F for false in each blanks.

(　　) 1. A precipitation reaction is similar to an acid-base titration and a complexometric titration in which these are often performed as volumetric methods.

(　　) 2. These titration methods make use of same types of reactions.

(　　) 3. A precipitation titration is a titration method in which the reaction of a titrant with a sample leads to an soluble precipitate.

(　　) 4. Most precipitation titrations that are used today are based on either the analysis of silver or the use of Ag^+ as a precipitating agent.

(   ) 5. There are the Volhard method, the Mohr method and the Fajans method in precipitation titrations.

C. Translate following sentences into Chinese.

1. A precipitation titration is generally carried out by adding to a sample a known volume of a solution containing a precipitating agent until no further precipitate is formed.

2. When the lack of further precipitation or a related signal indicates that the end point has been reached, the delivered volume of titrant is used to determine the concentration of the analyte in the original sample.

Reading Material 阅读材料

Precipitation Reactions

Precipitation reactions can be used to test for both anions and cations. For example, the presence of chloride ions in solution can be detected by adding silver ions. Silver ions attract chloride ions so strongly that if there are any chloride ions present, a white precipitate of silver chloride will form. Equally, silver ions can be tested for using chloride ions.

Many metals form insoluble hydroxides. So if a solution of sodium hydroxide is added to a solution of the metal, a precipitate of the hydroxide will be seen. Hydroxide precipitates vary in colors, and some redissolve when excess sodium hydroxide is added.

New Words

anion [ˈænaɪən] n. 阴离子
cation [ˈkætaɪən] n. 阳离子
presence [ˈprezns] n. 存在
detect [dɪˈtekt] vt. 探测，测定
redissolve [ˈriːdɪˈzɒlv] v. 再溶解

Expressions and Technical Terms

chloride ion 氯离子
silver chloride 氯化银
sodium hydroxide 氢氧化钠
vary in color 颜色变化

Exercises

A. Translate the following into English.

1. 阴离子　　　2. 阳离子　　　3. 探测　　　4. 氯离子

5. 氯化银　　　6. 金属　　　　7. 氢氧化钠　　8. 颜色变化

B. Choose the best answer to each question.

1. When you add some silver ions to the solution of chloride ions, what happens？（　　）

（A）A white precipitate will be seen at the bottom of the solution.

（B）You can not see any precipitates.

2. If a solution of sodium hydroxide is added to a solution of the metal, which can form insoluble hydroxide？（　　）

（A）An acid-base reaction.

（B）A precipitation reaction.

C. Cloze.

Precipitation reactions can be used to test for both _____ and _____ . For example, the presence of chloride ions in solution can be detected by adding _____ ions, and a white _____ of silver chloride will form. Equally, silver ions can be tested for using _____ ions.

Lesson 27
Redox Titrations 氧化还原滴定法

Warming up

氧化还原滴定法： 以溶液中氧化剂和还原剂之间的电子转移为基础的一种滴定分析方法。

试剂判断： 氧化还原不可分，得失电子是根本。失电子者被氧化，得电子者被还原。失电子者还原剂，得电子者氧化剂。

氧化还原滴定法应用非常广泛，它不仅可用于无机分析，而且可以广泛用于有机分析，许多具有氧化性或还原性的有机化合物可以用氧化还原滴定法来加以测定。

终点判断： 氧化还原滴定的化学计量点可借助仪器（如电位分析法）来确定，但通常借助指示剂来判断。有些滴定剂溶液或被滴定物质本身有足够深的颜色，如果反应后褪色，则其本身就可起指示剂的作用，例如高锰酸钾。可溶性淀粉遇痕量碘能显深蓝色，当碘被还原成碘离子时，深蓝色消失，因此在碘量法中，通常用淀粉溶液作指示剂。本身发生氧化还原反应的指示剂，例如二苯胺磺酸钠、次甲基蓝等，则在滴定到达化学计量点附近时，它也发生氧化还原反应，且其氧化态和还原态的颜色有明显差别，从而指示出滴定终点。

氧化还原反应的反应机理往往比较复杂，常伴随多种副反应，而且反应速率较低，有时需要加热或加催化剂来加速。这些干扰都需针对具体情况，采用不同的方法加以克服，否则会影响滴定的定量关系。

Text

A redox titration (more formerly known as an oxidation-reduction titration) is a titration method that makes use of an oxidation-reduction reaction. In the case of a COD analysis, the titrant is a reducing agent (Fe^{2+}) and the analyte (excess dichromate) is an oxidizing agent. The system has many of the same components used in other titration, such as a buret to deliver titrant to the sample and a means for detecting the end point. The end point in a redox titration can be detected either visually or by an instrumental method (e.g., by making potential measurements).

Like other type of titrations, a successful redox titration is based on a reaction that has a known stoichiometry between the analyte and titrant, a large equilibrium constant, and a fast reaction rate. The rates of oxidation-reduction reactions are more difficult to predict, but must be fast to result in an accurate end point. This is one reason why a back titration is used during COD analysis.

Two titrants for redox titrations that must be standardized using other chemicals are permanganate (MnO_4^-) and thiosulfate ($S_2O_3^{2-}$). Neither of these reagents can be made up directly as a usable titrant because their solid forms have insufficient purity and stability to act as primary standards. Dichromate is another common reagent that is used in redox titrations. It can be used directly as a titrant in a redox titration. Another important group of redox titrations involve the use of iodine as a titrant, reagent or analyte. This type of method is called an iodimetric titration (or iodimetry).

New Words

formerly [ˈfɔːməlɪ] adv. 原来，原先
dichromate [daɪˈkrəʊmeɪt] n. 重铬酸盐
component [kəmˈpəʊnənt] n. 组成部分，成分
visually [ˈvɪʒʊəlɪ] adv. 视觉上，外表上，看得见的
predict [prɪˈdɪkt] v. 预言，预测
permanganate [pəˈmæŋgəneɪt] n. 高锰酸盐
thiosulfate [ˌθaɪəʊˈsʌlfeɪt] n. 硫代硫酸盐
insufficient [ˌɪnsəˈfɪʃ(ə)nt] adj. 不足的，不够的
iodine [ˈaɪədiːn] n. 碘
iodimetry [aɪəʊˈdɪmɪtrɪ] n. 碘量法，碘量滴定法

Expressions and Technical Terms

redox titration 氧化还原滴定法
oxidation-reduction titration 氧化还原滴定法
reducing agent 还原剂
oxidizing agent 氧化剂
instrumental method 仪器分析法
potential measurement 电位测量
back titration 返滴定
primary standard 基准物
iodimetric titration 碘量法

Exercises

A. Translate the following into Chinese or English.
1. 氧化还原滴定法　　2. 氧化剂　　3. 还原剂　　4. 返滴定

5. 碘量法　　6. permanganate　　7. thiosulfate　　8. potential measurement

B. Choose the best answer to each question.
1. A redox titration (more formerly known as an oxidation-reduction titration) is a titration method that makes use of an (　　) reaction.

(A) acid-base　　　　　　(B) oxidation-reduction
(C) precipitation　　　　　(D) complexometric

2. The end point in a redox titration can be detected either (　) or by an (　).

(A) visually　　　　　　(B) instrumental method

3. A back titration is used during COD analysis, because the (　) of oxidation-reduction reactions are more difficult to predict.

(A) rate　　　　　(B) temperature　　　　　(C) concentration

C. Translate following sentences into Chinese.

1. A successful redox titration is based on a reaction that has a known stoichiometry between the analyte and titrant, a large equilibrium constant, and a fast reaction rate.

2. Neither of two titrants for redox titrations permanganate (MnO_4^-) and thiosulfate ($S_2O_3^{2-}$) can be made up directly as a usable titrant because their solid forms have insufficient purity and stability to act as primary standards.

Unit 2
Instrumental Analysis
仪器分析

Lesson 28
Gas Chromatography
气相色谱法

Warming up

气相色谱法： 是一种以气体为流动相的柱色谱法，根据所用固定相状态的不同可分为气固色谱（GSC）和气液色谱（GLC）。气相色谱法是20世纪50年代出现的一项重大科学技术成就。这是一种分离和分析技术，它在工业、农业、国防建设、科学研究中都得到了广泛应用。

原理： 气相色谱法的流动相为惰性气体，气固色谱法中以表面积大且具有一定活性的吸附剂作为固定相。当多组分的混合样品进入色谱柱后，由于吸附剂对每个组分的吸附力不同，经过一定时间后，各组分在色谱柱中的运行速度也就不同。吸附力弱的组分容易被解吸下来，最先离开色谱柱进入检测器，而吸附力最强的组分最不容易被解吸下来，因此最后离开色谱柱。如此，各组分得以在色谱柱中彼此分离，顺序进入检测器中被检测、记录下来。

根据色谱流出曲线上得到的每个峰的保留时间，可以进行定性分析，根据峰面积或峰高的大小，可以进行定量分析。

气相色谱仪 由以下五大系统组成：气路系统、进样系统、分离系统、温控系统、检测记录系统。组分能否分开，关键在于色谱柱；分离后组分能否鉴定出来则在于检测器，所以分离系统和检测系统是仪器的核心。

由于样品在气相中传递速度快，因此样品组分在流动相和固定相之间可以瞬间地达到平衡。另外加上可选作固定相的物质很多，因此气相色谱法是一个分析速度快和分离效率高的分离分析方法。近年来采用灵敏度高和选择性强的检测器，使其又具有分析灵敏度高、应用范围广等优点。

Text

In analytical chemistry, gas chromatography is a technique for separating chemical substances. Because of its simplicity, sensitivity, and effectiveness, gas chromatography is one of the most important tools in chemistry. It is widely used for purification of compounds and for the determination of some chemical constants and etc.

In the gas chromatography, we first introduce the test mixture or sample into an inert gas, the carrier stream. Liquid samples should be vaporized before injection into the carrier stream. Then the gas stream is passed through the packed column in which there is a stationary phase. The components of the sample having the greater interaction with the stationary phase are retarded to a greater extent. The components that have a lower interaction travel through the column at a faster rate. So sample components separate from each other. As each component leaves the column, it passes through a detector and then goes to a collector.

New Words

chromatography [ˌkrəʊməˈtɒɡrəfɪ] n. 色谱法
simplicity [sɪmˈplɪsətɪ] n. 简单，简易
determination [dɪˌtɜːmɪˈneɪʃ(ə)n] n. 测量，测定
sample [ˈsæmpl] n. 样品
injection [ɪnˈdʒekʃn] n. 注射
column [ˈkɒləm] n. 圆柱，柱状物
stationary [ˈsteɪʃənrɪ] adj. 固定的
component [kəmˈpəʊnənt] n. 成分
retard [rɪˈtɑːd] vt. 延迟，阻碍
detector [dɪˈtektə(r)] n. 探测器，检测器
collector [kəˈlektə(r)] n. 收集器

Expressions and Technical Terms

gas chromatography 气相色谱
analytical chemistry 分析化学
inert gas 惰性气体
carrier stream 载体流
packed column 填料柱
stationary phase 固定相

Exercises

A. Translate the following into English.

1. 气相色谱   2. 惰性气体   3. 收集器   4. 检测器

5. 固定相   6. 流动相   7. 纯化   8. 分析化学

B. Cloze.

In the gas chromatography, we first introduce the test _____ or sample into an inert gas, the _____ stream. Liquid samples should be _____ before injection into the carrier stream. Then the gas stream is passed through the _____ . The components of the sample having the _____ interaction with the stationary phase are retarded to a greater extent. The components that have a _____ interaction travel through the column at a faster rate. So sample components _____ from each other. As each component leaves the column, it passes through a _____ and then goes to a _____ .

C. Translate following sentences into Chinese.

1. Because of its simplicity, sensitivity, and effectiveness, gas chromatography is one of the most important tools in chemistry.

2. Gas chromatography is widely used for purification of compounds and for the determination of some chemical constants and etc.

Lesson 29
Atomic Absorption Spectrometry
原子吸收光谱法

Warming up

原子吸收光谱法（AAS）： 是基于待测元素的基态原子蒸气对其特征谱线的吸收，由特征谱线的特征性和谱线被减弱的程度对待测元素进行定性定量分析的一种仪器分析的方法。

由于原子能级是量子化的，因此在所有的情况下，原子对辐射的吸收都是有选择性的。由于各元素的原子结构和外层电子的排布不同，元素从基态跃迁至第一激发态时吸收的能量不同，因而各元素的共振吸收线具有不同的特征。由此可作为元素定性的依据，而吸收辐射的强度可作为定量的依据。AAS 现已成为无机元素定量分析应用最广泛的一种分析方法。该法主要适用样品中微量及痕量组分分析。

原子吸收光谱仪： 由光源、原子化系统、分光系统、检测系统等几部分组成。通常有单光束型和双光束型两类。这种仪器光路系统结构简单，有较高的灵敏度，价格较低，便于推广，能满足日常分析工作的要求；但其最大的缺点是不能消除光源波动所引起的基线漂移，对测定的精密度和准确度有一定的影响。

原子吸收光谱法已成为实验室的常规方法，能分析 70 多种元素，广泛应用于石油化工、环境卫生、冶金矿山、材料、地质、食品、医药等各个领域中。

原子吸收光谱法具有检出限低（火焰法可达 $\mu g/cm^3$ 级），准确度高（火焰法相对误差小于 1%），选择性好（即干扰少），分析速度快，应用范围广（火焰法可分析 70 多种元素，石墨炉法可分析 70 多种元素，氢化物发生法可分析 11 种元素）等优点。

Text

Atomic absorption spectrometry （AAS） is one of the most important techniques for the analysis and characterization of the elemental composition of materials and samples. Among the most common techniques for elemental analysis are flame emission, atomic absorption, and atomic fluorescence spectrometry.

All these techniques are based on the radiant emission, absorption, and fluorescence of atomic vapor. The key component of any atomic spectrometric method is the system for generating the atomic vapor （gaseous free atom or ion） from a sample, that is, the source. The most widely used sources are the flame and electrothermal atomizers.

The choice and development of an appropriate source for an atomic absorption measurement is the key step in the emergence of atomic absorption spectrometry as an analytical method. The primary source is the hollow-cathode lamp, which provides an almost ideal source for the atomic absorption measurement. Its wavelength exactly matches to that of analyte and the bandwidth is essentially ideal for the atoms in the flame.

New Words

spectrometry [spek'trɒmɪtrɪ] *n.* 光谱测定法
technique [tek'niːk] *n.* 技术，技能
composition [ˌkɒmpə'zɪʃ(ə)n] *n.* 组成
emission [i'mɪʃn] *n.* 发出，排放（物）
fluorescence [flə'resns] *n.* 荧光，荧光性
radiant ['reɪdɪənt] *adj.* 照耀的，辐射的
vapor ['veɪpə(r)] *n.* 蒸气，水蒸气
generate ['dʒenəreɪt] *vt.* 形成，造成，引起
atomizer ['ætəmaɪzə(r)] *n.* 喷雾器
appropriate [ə'prəʊprɪət] *adj.* 适当的，合适的
emergence [i'mɜːdʒəns] *n.* 出现，兴起
primary ['praɪmərɪ] *adj.* 首要的，初级的
wavelength ['weɪvleŋθ] *n.* 波长，波段
bandwidth ['bændwɪdθ] *n.* 带宽
essentially [ɪ'senʃəlɪ] *adv.* 本质上，根本上，本来

Expressions and Technical Terms

atomic absorption spectrometry (AAS) 原子吸收光谱法
elemental analysis 元素分析
flame emission 火焰发射
atomic fluorescence spectrometry 原子荧光光谱法
radiant emission 辐射发射
atomic vapor 原子蒸气
electrothermal atomizer 电热喷雾器
analytical method 分析方法
the hollow-cathode lamp 空心阴极灯

Exercises

A. Translate the following into English.
1. 原子吸收光谱　　2. 波长　　3. 带宽　　4. 元素分析

5. 火焰发射　　　　6. 原子荧光光谱法　　7. 分析方法　　8. 空心阴极灯

B. Choose the best answer to each question.

1. The key component of any atomic spectrometric method is the system for generating (　　) from a sample.
　　(A) the atomic vapor　　　　(B) electron

2. The choice and development of an appropriate source for an atomic absorption measurement is (　　) step in the emergence of atomic absorption spectrometry as an analytical method.
　　(A) the key　　　　(B) the first

3. The wavelength of the hollow-cathode lamp exactly matches to that of analyte and the (　　) is essentially ideal for the atoms in the flame.
　　(A) temperature　　　　(B) bandwidth

C. Translate following sentences into Chinese.

1. Atomic absorption spectrometry (AAS) is one of the most important techniques for the analysis and characterization of the elemental composition of materials and samples.

2. Among the most common techniques for elemental analysis are flame emission, atomic absorption, and atomic fluorescence spectrometry.

3. The primary source is the hollow-cathode lamp, which provides an almost ideal source for the atomic absorption measurement.

Lesson 30
Liquid Chromatography
液相色谱法

Warming up

液相色谱法是一类分离与分析技术，其特点是以液体作为流动相，固定相可以有多种形式，如纸、薄板和填充床等。

在色谱技术发展的过程中，为了区分各种方法，根据固定相的形式产生了各自的命名，如纸色谱、薄层色谱等。经典液相色谱法的流动相是依靠重力缓慢地流过色谱柱，因此固定相的粒度不可能太小（100～150μm）。分离后的样品是被分级收集后再进行分析的，使得经典液相色谱法不仅分离效率低、分析速度慢，而且操作也比较复杂。

高效液相色谱法（High Performance Liquid Chromatography，HPLC）是在经典液相色谱法的基础上，于20世纪60年代后期引入了气相色谱理论而迅速发展起来的。与经典液相色谱法的区别是填料颗粒小而均匀。因为较小的填充颗粒具有高柱效，但会引起高阻力，需用高压输送流动相，故又称高压液相色谱法。

液相色谱按其分离机理，可分为四种类型：吸附色谱、分配色谱、离子交换色谱、凝胶色谱。

高效液相色谱法应用非常广泛，几乎遍及定量定性分析的各个领域。

Text

For liquid chromatography, the procedure can be performed either in a column or on a plane. Columnar liquid chromatography is used for qualitative and quantitative analysis in a manner similar to the way in which gas chromatography is employed. For chemical analysis, the most popular category of columnar liquid chromatography is high performance liquid chromatography (HPLC). The method uses a pump to force one or more mobile phase solvents through high efficient, tightly packed columns. As with gas chromatography, an injection system is used to insert the sample into the entrance to the column, and a detector at the end of the column monitors the separated analyte components.

The stationary phase that is used for plane chromatography is held in place on a plane. Typically the stationary phase is attached to a plastic, metallic, or glass plate. Occasionally, a sheet of high quality filter paper is used as the stationary phase. The sample is added as a spot or a thin strip at one end of the plane. The mobile phase flows over the spot by capillary action during ascending development or as a result of the force gravity during descending development. During ascending development, the end of the plane near and below the sample is dipped into the mobile phase, and the mobile phase moves up and through the spot. During descending development, the mobile phase is added to the top of the plane and flows downward through the spot.

  Qualitative analysis is performed by comparing the retardation factors of the analyte components with the retardation factors of known substances. The retardation factor is defined as the distance from the original sample spot that the component has moved divided by the distance. The distance is that the mobile phase front has moved and is constant for a solute in a given solvent. Quantitative analysis is performed by measuring the sizes of the developed spots, by measuring some physical properties of the spots, or by moving the spots from the plane and assaying them by another procedure.

New Words

procedure [prəˈsiːdʒə(r)] n. 程序，手续，工序，过程，步骤
perform [pəˈfɔːm] v. 执行，履行，表演，扮演
plane [pleɪn] n. 水平，平面
columnar [kəˈlʌmnə(r)] adj. 柱形的，筒形的
qualitative [ˈkwɒlɪtətɪv] adj. 定性的，定质的
quantitative [ˈkwɒntɪtətɪv] adj. 定量的，数量（上）的
manner [ˈmænə(r)] n. 方式，方法，做法，态度，样子
category [ˈkætəgəri] n. 类型，部门，种类，类目
pump [pʌmp] n. 泵；v. 用泵抽，注入
plate [pleɪt] n. 盘子，盆子，平板
capillary [kəˈpɪləri] n. 毛细管，毛细血管，微管
ascend [əˈsend] v. 攀登，上升，爬坡
gravity [ˈɡrævəti] n. 重力，万有引力，重要性
descend [dɪˈsend] v. 下来，下降，下斜
constant [ˈkɒnstənt] n. 常数；adj. 不断的，持续的，永恒的
assay [əˈseɪ] n. 化验，试验

Expressions and Technical Terms

liquid chromatography 液相色谱法
stationary phase 固定相
mobile phase 流动相
flow over 流过
qualitative analysis 定性分析
analyte component 分析物成分

Exercises

A. Translate the following into English.

1. 液相色谱　　　2. 定性分析　　　3. 定量分析　　　4. 固定相

5. 流动相　　　　6. 攀爬　　　　　7. 下降　　　　　8. 重力

B. Decide whether the following statements are true (T) or false (F). Write T for true and F for false in each blank.

(　　) 1. Columnar liquid chromatography is used for qualitative and quantitative analysis in a manner similar to the way in which gas chromatography is employed.

(　　) 2. Liquid chromatography uses a pump to force one or more mobile phase solvents through high efficient, tightly packed columns.

(　　) 3. The sample is added as a spot or a thin strip at one end of the plane.

(　　) 4. During ascending development, the top of the plane is dipped into the mobile phase.

(　　) 5. During descending development, the mobile phase is added to the end of the plane and flows downward through the spot.

C. Comprehension exercises.

1. What are the difference and the similarity between gas chromatography and liquid chromatography?

2. How are the qualitative analysis and quantitative analysis performed in liquid chromatography?

译　文

液相色谱法

对于液相色谱法而言，该程序可在柱子中或平板上开展。柱式液相色谱用于定性或定量分析，与气相色谱中使用的方法相似。对于化学分析来说，最普遍的柱式液相色谱是高效液相色谱。该方法是使用泵使一种或多种流动相溶剂穿过高效的、紧密的填料柱。像气相色谱一样，使用注射系统将样品注入填料柱的入口，在柱子的末端用检测器检测被分离的组分。

薄层色谱中的固定相被固定在平板上。通常固定相被附着到塑料、金属或玻璃的平板上。偶尔，一层高质量的滤纸也被用作固定相。样品以点滴或细线的形式被加到平板的一端。流动相在上行展开法中通过毛细作用流过加样点；流动相在下行展开法中由于重力的作用流过加样点。在上行展开法中，接近且在加样点下方的平板一端被浸入到流动相中，流动相向上移动通过加样点。在下行展开法中，流动相被加入平板的顶端向下流动并通过加样点。

  通过比较分析物组分和已知物质的保留因子来进行定性分析。保留因子定义为组分移动的距离除以流动相前缘移动的距离，对于给定溶剂中的溶质而言，该数值是恒定的。通过测量斑点的大小或测量斑点的物理性质来进行定量分析，或者将斑点从平板上移动下来，用其他方法对其进行试验来进行定量分析。

Lesson 31
Ultraviolet and Visible Spectrophotometer
紫外-可见分光光度计

Warming up

紫外-可见分光光度计是基于紫外-可见分光光度法原理,利用物质分子对紫外-可见光谱区的辐射吸收来进行分析的一种分析仪器。主要由光源、单色器、吸收池、检测器和信号处理器等部件组成。

光源的功能是提供足够强度的、稳定的连续光谱。紫外光区通常用氢灯或氘灯。可见光区通常用钨灯或卤钨灯。

单色器的功能是将光源发出的复合光分解并从中分出所需波长的单色光。色散元件有棱镜和光栅两种。

吸收池: 又称比色皿,供盛放试液进行吸光度测量之用,其底及两侧为磨砂玻璃,另两面为光学透光面,为减少光的反射损失,吸收池的光学面必须完全垂直于光束方向。根据材质可分为玻璃池和石英池两种,前者用于可见光光区测定,后者用于紫外光区。

检测器: 是将光信号转变为电信号的装置,测量吸光度时,并非直接测量透过吸收池的光强度,而是将光强度转换为电流信号进行测试,这种光电转换器件称为检测器。常用的光电转换元件有光电管、光电倍增管及光电二极管阵列检测器。

信号显示系统: 是将检测器输出的信号放大,并显示出来的装置。

分光光度计的分类方法: 按光路系统可分为单光束和双光束分光光度计;按测量方式可分为单波长和双波长分光光度计;按绘制光谱图的检测方式分为分光扫描检测与二极管阵列全谱检测。

Text

Photometric methods are the frequently used of all spectroscopic methods. Ultraviolet and visible spectrophotometry (UV/Vis) is a powerful tool for quantitative analysis of samples free from turbidity in chemical research, biochemistry, chemical analysis, and industrial processing.

The amount of visible or other radiant energies absorbed by a solution is measured; since it depends on the concentration of the absorbing substance, it is possible to determine quantitatively the amount present. Colorimetric methods are based on the comparison of a colored solution of unknown concentration with one or more colored solutions of known concentration. In spectrophotometric methods, the ratio of the intensities of the incident and the transmitted beams of light is measured at a specific wavelength by means of a detector such as a photocell.

The absorption spectrum also provides a "fingerprint" for qualitatively identifying the absorbing substance, since the shape and intensity of UV/Vis absorption bands are related to electronic structure of the absorbing species. The molecule is often dissolved in a solvent to acquire the spectrum. Unfortunately, the spectra are often broad and frequently without fine structure. For this reason, UV absorption is much less useful for the qualitative identification of functional groups or particular molecules than analytical methods such as MS (mass spectrometry), IR (infrared spectroscopy), and NMR (nuclear magnetic resonance).

The measurement of absorption of ultraviolet-visible radiation is of a relative nature. One must continually compare the absorption of the sample with that of an analytical reference or a blank to ensure the reliability of the measurement.

The rate at which the sample and reference are compared depends on the design of the instrument. In single-beam instruments, there is only one light beam or optical path from the source to the detector. Thus, there is usually an interval of several seconds between measurements. Alternatively, the sample and reference may be compared many times a second, as in double-beam instruments. There are two main advantages of double-beam operation over single-beam operation. Very rapid monitoring of sample and reference help to eliminate errors due to drift in source intensity, electronic instability, and any changes in the optical system.

New Words

ultraviolet [ˌʌltrəˈvaɪələt] *adj.* 紫外的，紫外线的
visible [ˈvɪzəb(ə)l] *adj.* 可见的，明显的
spectrophotometer [ˌspektrəʊfəˈtɒmɪtə(r)] *n.* 分光光度计
photometric [ˌfəʊtəˈmetrɪk] *adj.* 光度计的，光度测定的
frequently [ˈfriːkwəntlɪ] *adv.* 频繁地，经常
spectroscopic [ˌspektrəˈskɒpɪk] *adj.* 分光镜的
turbidity [tɜːˈbɪdətɪ] *n.* 混浊，混乱，混浊度，浊度
radiant [ˈreɪdɪənt] *adj.* 照耀的，辐射的
determine [dɪˈtɜːmɪn] *v.* 查明，测定，准确算出，决定，下决心
comparison [kəmˈpærɪsn] *n.* 比较，对照，比喻
ratio [ˈreɪʃɪəʊ] *n.* 比，比率，比例，系数
intensity [ɪnˈtensətɪ] *n.* 强烈，强度
incident [ˈɪnsɪdənt] *n.* 事件，事变；*adj.* 入射的
wavelength [ˈweɪvleŋθ] *n.* 波长，波段
detector [dɪˈtektə(r)] *n.* 探测器，检测器
photocell [ˈfəʊtəʊsel] *n.* 光电池，光电管
identify [aɪˈdentɪfaɪ] *v.* 识别，确定，认同
spectrum [ˈspektrəm] *n.* 光谱，波谱，范围，系列

band [bænd] *n.* 范围，波段
acquire [əˈkwaɪə(r)] *v.* 获得，得到
spectra [ˈspektrə] *n.* 范围，光谱（spectrum 的复数形式），波谱
broad [brɔːd] *adj.* 宽的，广泛的，辽阔的，普遍的
optical [ˈɒptɪk(ə)l] *adj.* 视觉的，光学的
interval [ˈɪntəvl] *n.* 间隔
alternatively [ɔːlˈtɜːnətɪvlɪ] *adv.* 二者择一地，或者
eliminate [ɪˈlɪmɪneɪt] *v.* 排除，消灭
drift [drɪft] *n.* 漂移，偏移

Expressions and Technical Terms

colorimetric method 比色法
spectrophotometric method 光谱分析法
transmitted beam of light 透射光束
absorption spectrum 吸收光谱
fine structure 精细结构
ultraviolet-visible radiation 紫外可见辐射
analytical reference 分析性的参比
single-beam instrument 单光束仪器
double-beam instrument 双光束仪器
source intensity 光源强度
electronic instability 电子的不稳定性
optical system 光学系统

Exercises

A. Translate the following into English.
1. 紫外-可见分光光度计　　2. 波长　　3. 光谱　　4. 比色法

5. 波段　　6. 偏移　　7. 吸收光谱　　8. 检测器

9. 单光束分光光度计　　10. 双光束分光光度计　　11. 光学系统

B. Decide whether the following statements are true (T) or false (F). Write T for true and F for false in each blank.

(　　) 1. Ultraviolet and visible spectrometry (UV/Vis) is a powerful tool for quantitative analysis of samples.

(　　) 2. The amount of visible or other radiant energies absorbed by a solution is measured.

(　　) 3. The spectra are often broad and frequently without fine structure, so UV absorption is much less useful for the qualitative identification.

(　　) 4. The rate at which the sample and reference are compared depends on the design of the instrument.

(   ) 5. There are three main advantages of double-beam operation over single-beam operation.

C. Cloze.

1. Colorimetric methods are based on the comparison of a colored solution of unknown concentration with one or more colored solutions of known _____.

2. The absorption spectrum also provides a "_____" for qualitatively identifying the absorbing substance since the shape and intensity of UV/Vis absorption _____ are related to electronic structure of the absorbing species.

3. One must continually compare the absorption of the sample with that of an analytical _____ or _____ to ensure the reliability of the measurement.

4. Very rapid monitoring of sample and reference help to eliminate _____ due to drift in source intensity, electronic _____, and any changes in the optical system.

Lesson 32
Infrared Spectrometer 红外光谱仪

Warming up

红外吸收光谱法（简称红外光谱法）：利用被测物质的分子对不同波长的红外辐射吸收程度不同而对物质进行分析的方法。红外吸收光谱是由分子不停地作振动和转动运动而产生的，范围在 4000~400cm^{-1} 区域，它能提供大量的分子结构信息，是有机物的指纹峰，是进行基团诊断和结构鉴定的重要工具。

特点：适用范围广、特征性强，除旋光异构体及长链烷烃同系物外，几乎没有两个化合物具有相同的红外光谱。

红外光谱仪是利用物质对不同波长的红外辐射的吸收特性，进行分子结构和化学组成分析的仪器。红外光谱仪通常由光源，单色器，探测器和计算机处理信息系统组成。根据分光装置的不同，分为色散型和干涉型。对色散型双光路光学零位平衡红外分光光度计而言，当样品吸收了一定频率的红外辐射后，分子的振动能级发生跃迁，透过的光束中相应频率的光被减弱，造成参比光路与样品光路相应辐射的强度差，从而得到所测样品的红外光谱。

傅里叶变换红外光谱仪被称为第三代红外光谱仪，利用迈克尔森干涉仪将两束光程差按一定速度变化的复色红外光相互干涉，形成干涉光，再与样品作用。探测器将得到的干涉信号送入到计算机进行傅里叶变换的数学处理，把干涉图还原成光谱图。

Text

If a molecule absorbs infrared light (which has a lower energy than visible or ultraviolet light), this absorption is based on a change in the energy due to vibrations or rotations that are occurring in the molecule. A spectroscopic method that uses infrared light to study or measure chemicals is called infrared spectroscopy (or "IR spectroscopy").

IR spectroscopy is most frequently employed for qualitative identification of nearly pure compounds. Because each compound gives several peaks, the groups of atoms we call "functional groups" have characteristic vibrational energies and characteristic IR absorption wavelengths that can be used in this process.

Infrared spectroscopy is a type of absorption spectroscopy. Therefore, a dispersive - type infrared spectrophotometer will have the same basic components as the instruments used for the study of absorption of ultraviolet and visible radiation, although the sources, detectors, and materials used for the fabrication of optical elements will be different.

Although high-quality dispersive instruments are now in use and will continue to be produced, the most important development in instrumentation for infrared spectroscopy has been increased accessibility of dedicated high-speed computers. It has led to the proliferation of Fourier transform infrared spectroscopy.

New Words

infrared [ˌɪnfrəˈred] adj. 红外线的，（设备，技术）使用红外的；n. 红外区，红外线，红外辐射
vibration [vaɪˈbreɪʃn] n. 摆动，（偏离平衡位置的）一次性往复振动
rotation [rəʊˈteɪʃn] n. 旋转，转动
technique [tekˈniːk] n. 技术，技能
fabrication [ˌfæbrɪˈkeɪʃn] n. 制造，捏造，虚构
high-quality [ˌhaɪˈkwɒlətɪ] adj. 高级的，高品质的
accessibility [əkˌsesəˈbɪlətɪ] n. 易接近，可到达
dedicate [ˈdedɪkeɪt] v. 献（身），致力
proliferation [prəˌlɪfəˈreɪʃn] n. 繁殖，增生，激增

Expressions and Technical Terms

infrared spectrometer 红外光谱仪
infrared spectroscopy 红外光谱法
functional group 官能团
absorption spectroscopy 吸收光谱法
dispersive-type infrared spectrophotometer 色散型红外分光光度计
fabrication of optical element 光学元件的制造
Fourier transform infrared spectroscopy 傅里叶变换红外光谱法

Exercises

A. Translate the following into English.

1. 红外光谱仪　　　　2. 官能团　　　　3. 红外光谱法

4. 吸收光谱法　　　　5. 傅里叶变换红外光谱法　　6. 高品质的

B. Choose the best answer to each question.

1. Infrared spectroscopy is a type of (　　) spectroscopy.
　（A）absorption　　　　　　（B）emission

2. A dispersive-type infrared spectrophotometer will have the same basic components as the instruments used for the study of (　　).
　（A）absorption of ultraviolet and visible radiation.
　（B）flame emission spectrometry.

3. The most important development in instrumentation for infrared spectroscopy has been

increased accessibility of dedicated high-speed (　　)

(A) computers.　　　　　　　　(B) bond types.

C. Translate following sentences into Chinese.

1. A spectroscopic method that uses infrared light to study or measure chemicals is called infrared spectroscopy (or "IR spectroscopy").

2. IR spectroscopy is most frequently employed for qualitative identification of nearly pure compounds, because the groups of atoms we call "functional groups" have characteristic vibrational energies and characteristic IR absorption wavelengths.

Part 4

Analysis and Inspection Technology

分析检验技术

Lesson 33

Steps in Analysis and Inspection
分析检验的流程

Warming up

化学分析检测的对象不同、种类繁多、成分复杂、来源不一，分析的项目和要求也不尽相同，但不论哪种对象和要求，都要按照一个共同的程序进行分析检测，化学分析检测的一般程序是：

样品采集 → 试样制备 → 干扰消除 → 分析检测 → 结果计算和分析报告

Text

The analytical operation comprises up to five steps: (1) definition of the problem, (2) sampling and preliminary treatment, (3) separation, (4) measurement, and (5) calculation and evaluation of data. All are important.

Sampling

Before an analysis can be performed, a sample of the material to be analyzed must be obtained that is representative of the whole. This part of the operation is critical, since the quality of the analysis can be no better than the degree to which the sample represents the body of material under study. The study of proper sampling procedures is therefore of major importance.

Preliminary Operations

The method selected may require a variety of preliminary operations before the actual measurement. These may include dissolution of the sample, separation of the sought-for component from substances that would interfere in the measurement and adjustment of conditions for the measurement itself. Sometimes preliminary operations are unnecessary, as in measuring the free hydrogen ion concentration in a sample of lake or river water. At other times, they may constitute the major part of the analytical process. For example, in determining the distribution of various hydrocarbons in a gasoline sample, we use gas chromatography, where the measurement is simple once separation is achieved.

Measurement

Procedures for the measurement of a sought-for substance may be classified according to several general types. (1) Chemical methods involve adding to the sample a reagent that reacts in a defined way with the component to be determined. Either the amount of reagent needed to complete the reaction or the amount of product obtained can then be measured. (2) Electrical methods involve measuring potential, current, conductivity, or other electrical properties. They can be related to the amount of sought-for substance, either directly or upon addition of a reagent. (3) Spectroscopic methods involve measuring the quantity of electromagnetic radiation absorbed or emitted by a substance. Radiation from the visible, ultraviolet, and infrared regions of the spectrum is most often measured for this purpose; high energy (gamma rays or X-rays) or low energy (radio frequency) radiation is also measured, though less often. (4) Other physical methods include the measurement of properties such as density, refractive index, optical rotation, emission of alpha or beta particles from atomic nuclei, magnetism, and rates of chemical reaction.

Calculation and Evaluation of Results

Following the measurement step the analyst must relate the data (amount of reagent required, instrument reading, or other) to the amount of sought-for substance. Of equal importance is the assessment of the data in terms of precision, or reproducibility, and accuracy, or closeness to the true value. A variety of statistical tests are available to answer questions about the reliability of a set of measurements, rejection of a suspect measurement, and so on. These tests also help one decide how many analyses of a sample are necessary to give the required level of confidence in the results.

New Words

comprise [kəmˈpraɪz] v. 包括，构成，组成
preliminary [prɪˈlɪmɪnərɪ] adj. 初步的，初级的，预备的
definition [ˌdefɪˈnɪʃn] n. 定义，释义，清晰（度）
calculation [ˌkælkjuˈleɪʃn] n. 计算，估算，算计
evaluation [ɪˌvæljuˈeɪʃn] n. 评估，评价
representative [ˌreprɪˈzentətɪv] adj. 典型的，有代表性的
critical [ˈkrɪtɪkl] adj. 关键的，严重的，批评的
available [əˈveɪləbl] adj. 可用的，可获得的，有空的
interfere [ˌɪntəˈfɪə(r)] v. 干涉，妨碍
precision [prɪˈsɪʒn] n. 精确度，准确（性），精确
dissolution [ˌdɪsəˈluːʃn] n. 溶解，融化
component [kəmˈpəʊnənt] n. 组成部分，成分
adjustment [əˈdʒʌstmənt] n. 调整，调节，转变
gasoline [ˈɡæsəliːn] n. 汽油
chromatography [ˌkrəʊməˈtɒɡrəfɪ] n. 色谱分析法，层析法
potential [pəˈtenʃl] adj. 潜在的；n. 潜力，可能性
current [ˈkʌrənt] adj. 当前的，流通的；n. 水流，电流
conductivity [ˌkɒndʌkˈtɪvətɪ] n. 传导性，传导率，电导率
spectroscopic [ˌspektrəˈskɒpɪk] adj. 分光镜的，借助分光镜的
electromagnetic [ɪˌlektrəʊmæɡˈnetɪk] adj. 电磁的
visible [ˈvɪzəbl] adj. 可见的，明显的

ultraviolet [ˌʌltrəˈvaɪələt] adj. 紫外的，紫外线的，产生紫外线的
infrared [ˌɪnfrəˈred] adj. 红外线的
spectrum [ˈspektrəm] n. 光谱，波谱，范围，系列
refractive [rɪˈfræktɪv] adj. 折射的
optical [ˈɒptɪk(ə)l] adj. 视觉的，视力的，眼睛的，光学的
rotation [rəʊˈteɪʃn] n. 旋转，转动，轮流，循环
nuclei [ˈnjuːklɪaɪ] n. 核心，核子，原子核（nucleus 的复数形式）
magnetism [ˈmæɡnətɪzəm] n. 磁性，磁力，磁学，吸引力
reproducibility [rɪprəˌdjuːsəˈbɪlɪti] n. 再现性
statistical [stəˈtɪstɪkl] adj. 统计的，统计学的
rejection [rɪˈdʒekʃn] n. （对提议、建议或请求的）拒绝接受，嫌弃，厌弃
suspect [səˈspekt] adj. 不可靠的，不可信的，可疑的
analyses [əˈnæləsiːz] n. 分析，分解（analysis 的复数形式）

Expressions and Technical Terms

sampling and preliminary treatment 采样和前处理
calculation and evaluation of data 数据的处理和评价
no better than 几乎等于，倒不如
the sought-for component 待测组分
preliminary operation 前处理操作
gas chromatography 气相色谱
optical rotation 旋光度
refractive index 折射率
the true value 真值
level of confidence 置信水平

Exercises

A. Translate the following into Chinese or English.

1. the equipment available
2. confidence level in the results
3. operation comprised up to four steps
4. ultraviolet spectrum
5. 采样操作
6. 气相色谱
7. 电磁辐射
8. 评价试验数据
9. 将干扰组分从样品中分离出来
10. 待测组分
11. 化学反应速率
12. 分析步骤

B. Decide whether the following statements are true (T) or false (F) according to the text. Write T for true and F for false in each blank.

(   ) 1. The solutions of all analytical problems follow the same basic pattern that may include five steps.

(   ) 2. Sampling operations needn't be critical.

(   ) 3. The sample often needs preliminary treatment before the actual measurement.

(   ) 4. Preliminary operations are not necessary in measuring the free hydrogen ion concentration in a sample of lake water.

(   ) 5. If the direct determination of sample fails, the separation of the analyte from the interfering component will not become necessary.

(   ) 6. The results of the test should be assessed according to the precision, reproducibility, accuracy and closeness to the true value.

C. Translate the following paragraph into Chinese.

For quantitative analysis, the amount of sample taken is usually measured by mass or volume. Where a homogeneous sample already exists, it may be subdivided without further treatment. With many solids such as ores, however, crushing and mixing are a prior requirement. The sample often needs additional preparation for analysis, such as drying, ignition and dissolution.

Lesson 34
Sample Preparation
试样的制备

Warming up

分析检验的首项工作就是如何从大批物料中（即总体中）采取符合分析工作要求的实验室样品（即样品）作为分析试样，这项工作称为**试样的采集与制备**。

采取的样品是否合适，制备是否得当，能不能代表整体样品的情况都是至关重要的；如果采样不当，则后续操作再正确，分析工作再认真都是徒劳的。

工业物料的数量，往往以千万吨计，其组成有的比较均匀，有的很不均匀。而我们对物料进行分析时所需的试样量是很少的，多不过数克，甚至更少，对这些少量试样的分析结果必须能代表全部物料的平均组成。因此，**正确采集和制备具有充分代表性的样品是采样的基本原则，也是分析检测结果是否准确的先决条件**。

样品的制备方法因产品类型和检测项目不同而异。在一般分析工作中，通常先要将待测试样进行预处理，目的是将固体试样处理成溶液，或将组成复杂的试样处理成简单、便于分离和测定的形式。常用的预处理方法有有机物破坏法、溶剂提取法、蒸馏法、化学分离法、色谱分离法、浓缩法等。试样分解是定量分析中很重要的步骤，常用的试样分解方法有酸碱分解法、熔融分解法、湿法消化和干法灰化等。

随着各种高精密高自动化的先进仪器不断出现，使得样品分析工作能够快速而准确地进行，然而由于样品的采集和制备等工作的复杂性和多样性，造成样品处理耗时费力、且引入污染多等，成为分析检验工作的瓶颈。

Text

Few samples in the real world can be analyzed without some chemical or physical preparation. The aim of all sample preparation is to provide the analyte of interest in the physical form required by the instrument, free of interfering substances, and in the concentration range required by instrument. For many instruments, a solution of analyte in organic solvent or water is required.

Solid samples may need to be crushed or ground, or they may need to be washed with water, acid, or solvent to remove surface contamination. Liquid samples with more than one phase may need to be extracted or separated. Filtration or centrifugation may be required.

The type of sample preparation needed depends on the nature of the sample, the analytical technique chosen, the analyte to be measured, and the problem to be solved. Most samples are not homogeneous. Many samples contain components that interfere with the determination of the analyte. A wide variety of approaches to sample preparation has been developed to deal with these problems in real samples.

Acid Dissolution and Digestion

Metals, alloys, ores, geological samples, and glass react with concentrated acids and this approach is commonly used for dissolving such samples. Organic materials can be decomposed (digested or "wet ashed") using concentrated acids to remove the carbonaceous material and solubilize the trace elements in samples, such as biological tissues, foods and plastics.

Fusions

Heating a finely powdered solid sample with a finely powdered salt at high temperatures, until the mixture melts is called a fusion or molten salt fusion. The reacted and cooled melt is leached with water or dilute acid to dissolve the analytes for determination of elements.

Dry Ashing and Combustion

To analyze organic compounds or substances for the inorganic elements present, it is often necessary to remove the organic material. Wet ashing with concentrated acids has been mentioned as one way of doing this. The other approach is "dry ashing", that is, ignition of the organic material in air or oxygen. The organic components react to form gaseous carbon dioxide and water vapor, leaving the inorganic components behind as solid oxides.

Extraction

If we want to determine organic analytes, the most common approach for organic analytes is to extract the analytes out of the sample matrix using a suitable solvent. Solvents are chosen with the polarity of the analyte in mind, since "like dissolves like". That is, polar solvents dissolve polar compounds, while nonpolar solvents dissolve nonpolar compounds.

New Words

preparation [ˌprepəˈreɪʃn] n. 准备（工作），制剂
analyst [ˈænəlɪst] n. 分析者，化验员
interfere [ˌɪntəˈfɪə(r)] v. 干涉，妨碍
analyte [ænəˈlaɪt] n. （被）分析物，分解物
crush [krʌʃ] v. 挤，捣碎
ground [graʊnd] adj. 磨细的，磨碎的
contamination [kənˌtæmɪˈneɪʃn] n. 污染，弄脏，毒害
extract [ˈekstrækt] n. 提取，提炼
filtration [fɪlˈtreɪʃn] n. 过滤，筛选
centrifugation [sentrɪfjʊˈgeɪʃən] n. 离心法，离心过滤
homogeneous [ˌhɒməˈdʒiːnɪəs] adj. 同性质的，同类的，均匀的
variety [vəˈraɪətɪ] n. 多种多样，多样化
dissolution [ˌdɪsəˈluːʃn] n. 溶解，融化

digestion [daɪˈdʒestʃən] *n.* 消化（系统）
ore [ɔː(r)] *n.* 矿，矿石，矿砂
fusion [ˈfjuːʒn] *n.* 融合，熔解，熔化
finely [ˈfaɪnli] *adv.* 美好地，精细地，细微地
melt [melt] *v.* （使）熔化，融化，与……融为一体
leach [liːtʃ] *v.* （将化学品、矿物质等）过滤，（液体）过滤，滤去
combustion [kəmˈbʌstʃən] *n.* 燃烧，烧毁，氧化
ignition [ɪgˈnɪʃn] *n.* 发火装置，着火，燃烧，点火，点燃
matrix [ˈmeɪtrɪks] *n.* 基质，母体
polarity [pəˈlærəti] *n.* 极性
polar [ˈpəʊlə(r)] *adj.* 极地的，磁极的，完全相反的
nonpolar [ˌnɒnˈpəʊlə(r)] *adj.* 无极的

Expressions and Technical Terms

sample preparation 样品的制备
deal with 解决，处理
wet ashed 湿法灰化
dry ashing 干法灰化
trace element 痕量组分
biological tissue 生物组织
molten salt fusion 熔盐熔融
solid oxide 固体氧化物
like dissolves like 相似相溶

Exercises

A. Translate the following into English.

1. 制样　　　　2. 破碎　　　　3. 研磨　　　　4. 过筛

5. 萃取　　　　6. 离心分离　　7. 均匀的　　　8. 酸溶法

9. 消化　　　　10. 熔融　　　　11. 灰化　　　　12. 相似相溶

B. Decide whether the following statements are true (T) or false (F) according to the text. Write T for true and F for false in each blank.

(　) 1. Many samples in the real world can be analyzed without some chemical or physical preparation.

(　) 2. Liquid samples with more than one phase may need to be extracted or separated.

(　) 3. Heating a finely powdered solid sample with a finely powdered salt at low temperatures, until the mixture melts is called a fusion or molten salt fusion.

(　) 4. By "dry ashing", the organic components react to form gaseous carbon dioxide and water vapor, leaving the inorganic components behind as solid oxides.

(　　) 5. If we want to determine organic analytes, the most common approach for organic analytes is to extract the analytes out of the sample matrix using a suitable solvent.

C. Choose the best answer to each question.

1. The aim of all sample preparation is to provide the analyst of interest (　　).
(A) in the physical form required by the instrument
(B) free of interfering substances
(C) in the concentration range required by instrument

2. Solid samples may need to be (　　) or ground, or they may need to be washed with water, acid, or solvent to remove surface contamination.
(A) crushed　　　(B) heated　　　(C) colored

3. Organic materials can be decomposed (digested or "wet ashed") using concentrated (　　) to remove the carbonaceous material.
(A) acids　　　(B) bases　　　(C) salts

4. To "wet ashing" with concentrated acids, the other approach with ignition of the organic material in air or oxygen is (　　).
(A) "dry ashing"　　　(B) digestion　　　(C) fusions

D. Translate following sentences into Chinese.

1. The aim of all sample preparation is to provide the analyte of interest in the physical form required by the instrument, free of interfering substances, and in the concentration range required by instrument.

2. The type of sample preparation needed depends on the nature of the sample, the analytical technique chosen, the analyte to be measured, and the problem to be solved.

3. "Like dissolves like". That is, polar solvents dissolve polar compounds, while nonpolar solvents dissolve nonpolar compounds.

Lesson 35
Organic Compound Identification Using Infrared Spectroscopy
运用红外光谱法鉴别有机物

Warming up

以共价键相连的原子就像用弹簧连接的球,它们在不停地以各种形式振动着。当电磁波辐射频率与基团振动的频率相同时,分子便吸收电磁波的能量而发生振动能级的跃迁,使振动的振幅加大。红外光谱仪记录下吸收峰的位置,便获得红外光谱图。

共价键的振动包括伸缩振动(stretching vibration)和弯曲振动(bending vibration)两种。量子化学理论证明,并不是任意两个能级间都能发生跃迁。若两个能级的跃迁根据光谱选律是"允许"的,出现较强的吸收峰;若跃迁是"禁阻"的,出现较弱的峰,甚至观察不到信号。

一般来说,不同基团有不同的红外光谱吸收,而同一基团在不同的化合物中,其吸收位置差异不大,因此通常可通过红外光谱确定某化合物中存在的键和官能团的种类。

红外光谱图中的横坐标代表波长,通常用波数 σ(即波长的倒数,$\sigma = 1/\lambda$)表示,单位为 cm^{-1},红外光谱的波数范围在 $4000 \sim 400 cm^{-1}$ 之间。纵坐标代表光透过程度,常用透过率 $T(\%)$ 表示。一般将整个红外光谱分为两个区,官能团区(functional group region)和指纹区(fingerprint region)吸收峰的位置与分子结构有关。

红外光谱看起来相当复杂,一般并不需要尝试去解析一个红外光谱图中所有的吸收峰,只需关心那些涉及某种化合物结构的特征吸收峰。当要确定一个未知化合物的结构时,仅凭红外光谱是不够的,通常还需要结合其他波谱,如核磁共振、质谱等结构分析信息。

红外光谱的吸收强度可定性地用很强(very strong,缩写为 vs)、强(strong,缩写为 s)、中强(medium,缩写为 m)、弱(weak,缩写为 w)、可变(variable,缩写为 v)等符号来表示。吸收峰的形状分为宽峰、尖峰、肩峰、双峰等类型。

红外光谱中一些主要化学键的伸缩振动波数

键的类型	σ/cm^{-1}	强度/吸收峰峰形
C≡N	$2260 \sim 2220$	中强

（续表）

键的类型	σ/cm^{-1}	强度/吸收峰峰形
C≡C	2260～2100	中强到弱
C=C	1680～1600	中强
C=N	1690～1640	中强
苯环	～1600 和 1500～1430	强到弱
C=O	1780～1650	强
C—O	1250～1050	强
C—N	1230～1020	中强
C—C	1200～800	较弱
O—H（醇）	3650～3580（游离）	尖
	3550～3450（二聚）	中强,较尖
	3400～3200（多聚）	强,宽
	3600～2500（分子内缔合）	宽,较弱
O—H（酸）	～3550（游离）	强,较尖
	3300～2500（二聚）	强,宽
N—H	3500～3300	中强
C—H	3300～2700	中强

Text

Infrared spectroscopy can be utilized in the qualitative examination of organic compounds even more fruitfully, as it gives information on the presence or absence of almost all functional groups.

A disadvantage of the technique is that aqueous solutions cannot, in general, be tested, and the instrument is more complex and expensive than a UV spectrophotometer. Further, the evaluation of IR spectra requires more practice and greater theoretical knowledge.

The most certain and unambiguous way of identifying an unknown pure compound is to compare its IR spectrum with a reference spectrum. Today several catalogues of a spectra are available.

When the pure sample is completely unknown and no comparison with a reference substance is possible, it is advisable to start with chemical methods, qualitative and quantitative elemental analysis and determination of molecular weight. After having established the empirical formula, the individual bond types and functional groups can be identified on the IR absorption bands.

The values of characteristic bond and group frequencies have been determined from the spectra of several different compounds with known structure. Although this empirical determination resulted in approximate values only, they can be applied very successfully in practice. The presence of individual functional groups may be confirmed by chemical analytical methods and their number in the molecule can be determined by quantitative functional group analysis.

When the intense and medium intensity bands have been assigned, the low-intensity bands are examined. This is a more difficult task, as the combination bands and overtones also appear with similar intensities and can hardly, if at all, be distinguished from weak fundamental bands.

In the evaluation of IR spectra it must be borne in mind that the positions of bands are altered by conjugation and the formation of associated structures and may also be affected by the preparation technique applied in recording the spectrum.

Spectra recorded on mixtures or contaminated samples can be evaluated only with difficulty and chemical or physical separation is necessary.

New Words

identification [aɪˌdentɪfɪˈkeɪʃn] n. 识别，鉴别，鉴定，辨识
utilize [ˈjuːtəlaɪz] v. 利用，使用
fruitfully [ˈfruːtfəlɪ] adv. 产量多地，肥沃地
disadvantage [dɪsədˈvɑːntɪdʒ] n. 不利条件，劣势
unambiguous [ˌʌnæmˈbɪɡjʊəs] adj. 不含糊的，清楚的
catalogue [ˈkætəlɒɡ] n. 目录，目录册
spectra [ˈspektrə] n. 光谱，范围（spectrum 的复数形式）
comparison [kəmˈpærɪs(ə)n] n. 比较，对照
empirical [ɪmˈpɪrɪk(ə)l] adj. 以经验为依据的
confirm [kənˈfɜːm] v. 确定，证实，认可
intense [ɪnˈtens] adj. 强烈的
assign [əˈsaɪn] v. 分派，确定
overtone [ˈəʊvətəʊn] n. 暗示，弦外之音
distinguish [dɪˈstɪŋɡwɪʃ] v. 使有别于，认出，区别
borne [bɔːn] v. 忍受（bear 的过去分词）
conjugation [ˌkɒndʒʊˈɡeɪʃ(ə)n] n. 结合，配合
contaminate [kənˈtæmɪneɪt] v. 污染，弄脏，腐蚀

Expressions and Technical Terms

organic compound 有机化合物
infrared spectroscopy 红外光谱法
qualitative examination 定性检验
functional group 官能团
UV spectrophotometer 紫外分光光度计
theoretical knowledge 理论知识
IR spectrum 红外光谱
empirical formula 经验式，实验式
molecular weight 分子量
IR absorption band 红外吸收谱带

group frequency 基团频率
approximate value 近似值
combination band 综合谱带
fundamental band 基本吸收带
associated structure 缔合结构

Exercises

A. Translate the following into English or Chinese.

1. 有机化合物　　　2. 官能团　　　3. 谱带　　　4. 分子量

5. infrared spectroscopy　　　6. UV spectrophotometer　　　7. empirical formula

8. qualitative examination　　　9. IR absorption band

B. Decide whether the following statements are true (T) or false (F) according to the text. Write T for true and F for false in each blank.

(　) 1. Infrared spectroscopy gives information on the presence or absence of all functional groups.

(　) 2. A disadvantage of the technique is that aqueous solutions cannot be tested generally.

(　) 3. The presence of individual functional groups may be confirmed by chemical analytical methods.

(　) 4. The low-intensity bands are examined easily.

(　) 5. The positions of bands are altered by conjugation and the formation of associated structures and may also be affected by the preparation technique applied in recording the spectrum.

C. Comprehension exercises.

1. What is the disadvantage of the infrared spectrum technique in the qualitative examination of organic compounds?

2. When the pure sample is completely unknown and no comparison with a reference substance is possible, what should you do?

3. What should we do before spectra recorded on mixtures or contaminated samples?

Lesson 36

Determination of Sulfur Dioxide in Ambient Air
环境空气中二氧化硫的测定

Warming up

大气污染： 通常是指由于人类活动和自然过程引起某种物质进入大气中，积累到足够的浓度、达到了足够的时间并因此危害了人体的健康或危害了环境的现象。

大气污染物的种类很多，目前引起人们注意的有 100 多种。

大气污染物的分类有很多方式，主要有：

（1）一次污染物和二次污染物　一次污染物是指直接从污染源排放的污染物质，如二氧化硫、二氧化氮、一氧化碳、颗粒物等；二次污染物是指由一次污染物在大气中互相作用，经化学反应或光化学反应形成的与一次污染物的物理、化学性质完全不同的新的大气污染物。最常见的二次污染物如硫酸及硫酸盐气溶胶、硝酸及硝酸盐气溶胶、臭氧、光化学烟雾等。

（2）天然污染物和人为污染物

（3）气态污染物和粒子状态污染物　二氧化硫（sulfur dioxide）是最常见、有毒、有刺激性的硫氧化物，化学式 SO_2，无色气体，大气主要污染物之一。当二氧化硫溶于水中，会形成亚硫酸。若亚硫酸进一步在细颗粒物即粒子状态污染物（如 PM2.5）存在的条件下氧化，便会迅速高效生成硫酸（酸雨的主要成分）。

世界卫生组织国际癌症研究机构公布的致癌物清单中二氧化硫在 3 类致癌物清单中。

目前，最常用的两种测量空气中二氧化硫浓度的方法：四氯汞钾溶液吸收盐酸副玫瑰苯胺分光光度法和甲醛溶液吸收盐酸副玫瑰苯胺分光光度法。

Text

When air contains one or more chemicals and harms humans, other animals, plants or materials, the phenomenon is called air pollution.

There are two major types of air pollutants. A primary air pollutant is a chemical added directly to the air and pollution occurs in a harmful concentration of it. A secondary air pollutant is formed in the atmosphere through a chemical reaction.

Major air pollutants are as following: (1) Carbon oxides—carbon monoxide (CO), carbon dioxide (CO_2); (2) Sulfur oxides—sulfur dioxide (SO_2), sulfur trioxide (SO_3); (3) Nitrogen oxides—nitrogen dioxide (NO_2), dinitrogen oxide (N_2O), nitrogen oxide (NO); (4) Hydrocarbons—methane (CH_4), butane (C_4H_{10}), benzene (C_6H_6); (5) Smoke, dust; (6) Inorganic compounds—hydrogen fluoride (HF), hydrogen sulfide (H_2S), ammonia (NH_3), sulfuric acid (H_2SO_4), nitric acid (HNO_3); (7) Noise.

Sulfur dioxide in ambient air is determined by formaldehyde solution sampling-pararosaniline hydrochloride spectrophotometry.

Principle: After sulfur dioxide is absorbed by formaldehyde buffer solution, a stable hydroxymethyl sulfonic acid addition-compound is formed. Sodium hydroxide is added to the sample solution to decompose the addition-compound. The released sulfur dioxide reacts with pararosaniline and formaldehyde to generate purple compound. The absorbance was measured at the wavelength of 577nm with a spectrophotometer.

Instruments and equipment: spectrophotometer, fritted glass bubbler, water bath, colorimetric tube, air sampler.

Method:

1. Sample collection and preservation.

2. The drawing of the calibration curve.

3. Determination: If there is any turbidity in the sample solution, it shall be removed by centrifugation. Samples were placed for 20min to decompose the ozone. To samples collected in a short time, move the sample solution in the absorption tube into a 10mL colorimetric tube, wash the absorption tube with a small amount of formaldehyde absorption liquid, merge the lotion into the colorimetric tube and dilute it to the marking line. Add 0.5mL of sodium ammonia sulfonate solution, mix, and place it for 10min to remove interference from nitrogen oxides. The following procedure is same to the calibration curve. The absorbance was measured at the wavelength of 577nm with a spectrophotometer.

4. Calculation of the concentration of sulfur dioxide in the air.

New Words

ambient [ˈæmbɪənt] *adj*. 环境的，周围的

pollutant [pəˈluːtənt] *n*. 污染物

harm [hɑːrm] *vt*. 伤害，损害

phenomenon [fəˈnɒmɪnən] *n*. 现象

primary [ˈpraɪmərɪ] *adj*. 主要的，初步的，初级的

concentration [ˌkɒns(ə)nˈtreɪʃ(ə)n] *n*. 浓度

emit [ɪˈmɪt] *vt*. 发出，散发

secondary [ˈsekənd(ə)rɪ] *adj*. 二级的，中级的

methane [ˈmeθeɪn] *n*. 甲烷，沼气

butane [ˈbjuːteɪn] *n*. 丁烷

benzene [ˈbenziːn] *n*. 苯

dust [dʌst] *n*. 灰尘，尘土

formaldehyde [fɔːˈmældɪhaɪd] *n*. 甲醛
pararosaniline [ˌpærərəʊˈzænɪˌliːn] *n*. 碱性副品红
principle [ˈprɪnsəpl] *n*. 准则，原理
hydroxymethyl [haɪdrɒksɪˈmeθɪl] *n*. 羟甲基
frit [frɪt] *vt*. 烧结
colorimetric [ˌkʌlərɪˈmetrɪk] *adj*. 比色的，色度的
turbidity [tɜːˈbɪdətɪ] *n*. 浊度，浑浊
centrifugation [sentrɪfjʊˈgeɪʃən] *n*. 离心分离
ozone [ˈəʊzəʊn] *n*. 臭氧
merge [mɜːdʒ] *vt*. （使）合并，（使）融合
lotion [ˈləʊʃn] *n*. 洗剂

Expressions and Technical Terms

air pollution 空气污染
carbon monoxide 一氧化碳
sulfur dioxide 二氧化硫
sulfur trioxide 三氧化硫
nitrogen dioxide 二氧化氮
dinitrogen oxide 一氧化二氮
inorganic compound 无机化合物
hydrogen fluoride 氟化氢
hydrogen sulfide 硫化氢
formaldehyde solution sampling-pararosaniline hydrochloride spectrophotometry
甲醛溶液吸收-盐酸副玫瑰苯胺分光光度法
buffer solution 缓冲溶液
sulfonic acid 磺酸
addition-compound 加成化合物
sodium hydroxide 氢氧化钠
fritted glass bubbler 多孔玻板吸收管
colorimetric tube 比色管
air sampler 空气采样器
calibration curve 标准曲线
absorption tube 吸收管
sodium ammonia sulfonate 氨磺酸钠

Exercises

A. Translate the following into English or Chinese.
1. 大气污染　　2. 一次大气污染物　　3. 二次大气污染物　　4. 碳氧化合物

5. 氮氧化合物　　6. 硫氧化合物　　7. 碳氢化合物　　8. 粉尘

9. 苯　　　　　10. 噪声　　　　11. 臭氧　　　　　12. 缓冲溶液

13. colorimetric tube　　14. calibration curve　　15. absorption tube

B. Cloze.

After _____ _____ is absorbed by formaldehyde _____ _____, a stable hydroxymethyl sulfonic acid addition-compound is formed. Sodium hydroxide is added to the sample solution to _____ the addition-compound. The released sulfur dioxide _____ with pararosaniline and formaldehyde to generate _____ compound. The absorbance was measured at the _____ of 577nm with a spectrophotometer.

C. Translation.

1. There are two major types of air pollutants: a primary air pollutant and a secondary air pollutant.

2. Major air pollutants are as following: carbon oxides, sulfur oxides, nitrogen oxides, hydrocarbons, smoke, dust, inorganic compounds, noise.

3. Sulfur dioxide in ambient air is determined by formaldehyde absorbing-pararosaniline spectrophotometry.

4. There are about four steps in determination of sulfur dioxide in ambient air: sample collection and preservation, the drawing of the calibration curve, determination and calculation.

Reading Material　阅读材料

The Atmosphere

The atmosphere is a mixture of several gases. There are about ten chemical elements which remain in gaseous form under all natural conditions. Of these gases, oxygen makes up about 21 percent and nitrogen about 78 percent. Several other gases, such as argon, carbon dioxide, hydrogen, neon, krypton and xenon, comprise the remaining one percent of the volume of dry air. The amount of water vapor is very important in weather changes.

The layer of the air next to the earth, which extends upward for about ten miles, is known as the troposphere. It makes up about 75 percent of all the weight of the atmosphere. It is the warmest part of the atmosphere because most of the solar radiation is absorbed by the earth's surface.

The upper layers are colder because of their greater distance from the earth's surface. It was assumed that upper air had little influence on weather changes. Recent studies have shown the assumption to be incorrect.

> 大气组成，也就是大气的化学成分，一般是指干空气（dry air）的化学成分。大气是多相态的体系，包含干空气、水和颗粒物，但是后两者含量太低且不稳定，所以一般说到大气组成，指的都是干空气的化学成分。
>
> 从地表到高度为 105 km 的大气层中，各种气体占干空气的体积分数如下：氮气（N_2）78.08%；氧气（O_2）20.95%；氩气（Ar）0.93%；二氧化碳（CO_2）380×10^{-6}；氖气（Ne）18×10^{-6}；氦气（He）5×10^{-6}；甲烷（CH_4）1.75×10^{-6}；氪气（Kr）1×10^{-6}；氢气（H_2）0.5×10^{-6}；一氧化二氮（N_2O）0.3×10^{-6}；臭氧（O_3）$0\sim0.1\times10^{-6}$。可以看出，氮气和氧气是大气中含量最高的气体，其次是氩气，它在惰性气体里是含量最高的。干空气中除了以上这些气体，还有极少量的其他气体。
>
> 除惰性气体外，各气体的浓度和停留时间呈正相关。

New Words

atmosphere [ˈætməsfɪə(r)] *n.* 大气，空气，大气层
natural [ˈnætʃ(ə)rəl] *adj.* 自然的，天生的
condition [kənˈdɪʃn] *n.* 条件，环境
argon [ˈɑːgɒn] *n.* 氩
neon [ˈniːɒn] *n.* 氖
krypton [ˈkrɪptɒn] *n.* 氪
xenon [ˈzenɒn] *n.* 氙
comprise [kəmˈpraɪz] *v.* 包含，由……组成
extend [ɪkˈstend] *v.* 扩充，延伸
troposphere [ˈtrɒpəsfɪə(r)] *n.* 对流层
solar [ˈsəʊlə(r)] *adj.* 太阳的，日光的
assume [əˈsjuːm] *vt.* 假定，设想
assumption [əˈsʌmpʃn] *n.* 假定，设想

Expressions and Technical Terms

make up 组成
water vapor 水蒸气
solar radiation 太阳辐射

Exercises

A. Translate the following into English.

1. 大气层　　　　2. 氮　　　　3. 氩　　　　4. 氙

5. 太阳辐射　　　6. 水蒸气　　　7. 吸收　　　8. 假定

B. Choose the best answer to each question.

1. The atmosphere consists of (　　)
（A）nine chemical elements.
（B）only nitrogen and oxygen.
（C）about 21 percent oxygen and about 78 percent nitrogen and other gases.

2. The troposphere is the warmest part of the atmosphere because (　　)
（A）most of the solar radiation is absorbed by the earth's surface.
（B）it is nearest the sun.
（C）it contains heat.

3. The upper layers are colder than the troposphere because of (　　)
（A）their most of the solar radiation.
（B）their greater distance from the earth's surface.

Lesson 37
Determination of Water Hardness
水硬度的测定

Warming up

硬度是水质的一个重要监测指标，通过监测可以知道被监测的水是否可用于工业生产及日常生活。高硬度的水可使洗涤剂的效用大大降低；纺织工业中硬度过大的水会使纺织物粗糙且难以染色；硬度高的水烧锅炉易堵塞管道，引起锅炉爆炸事故；高硬度的水难喝、有苦涩味，饮用后甚至影响胃肠功能等。因此水硬度的测定方法研究是不容忽视的。

水总硬度： 是指水中 Ca^{2+}、Mg^{2+} 的总量，它包括暂时硬度和永久硬度。

碳酸盐硬度（暂时硬度）： 主要是由钙、镁的碳酸氢盐[$Ca(HCO_3)_2$、$Mg(HCO_3)_2$]所形成的硬度，还有少量的碳酸盐形成的硬度。碳酸氢盐硬度经加热之后分解成沉淀物可从水中除去，故亦称为暂时硬度。

非碳酸盐硬度（永久硬度）： 主要是由钙、镁的硫酸盐、氯化物和硝酸盐等盐类所形成的硬度。这类硬度不能用加热分解的方法除去，故也称为永久硬度，如 $CaSO_4$、$MgSO_4$、$CaCl_2$、$MgCl_2$、$Ca(NO_3)_2$、$Mg(NO_3)_2$ 等。

钙硬度： 水中 Ca^{2+} 的含量称为钙硬度。**镁硬度：** 水中 Mg^{2+} 的含量称为镁硬度。

不同国家的换算单位也有不同的标准。各种水质硬度换算单位如下：

（1）德国度　1度相当于1L水中10mg的CaO；
（2）英国度　1度相当于0.7L水中10mg的$CaCO_3$；
（3）法国度　1度相当于1L水中10mg的$CaCO_3$；
（4）美国度　1度相当于1L水中1mg的$CaCO_3$。

分析测定硬度的方法很多，主要可分为**化学分析法**和**仪器分析法**。

（1）化学分析法：EDTA配位滴定法是一种普遍使用的测定水的硬度的化学分析方法。它是在一定条件下，以铬黑T为指示剂，pH=10的 $NH_3 \cdot H_2O\text{-}NH_4Cl$（氨-氯化铵）为缓冲溶液，EDTA与钙、镁离子形成稳定的配合物，从而测定水中钙镁离子总量。

（2）仪器分析法有分光光度法、原子吸收法、色谱分析法、电极法等。

Text

EDTA titrations are routinely used to determine water hardness in a laboratory. Raw well water samples can have a significant quantity of dissolved minerals which contribute to a variety of problems associated with the use of such water. These minerals consist chiefly of calcium and magnesium carbonates, sulfates, *etc.*, and occasionally some iron. The problems that arise are mostly a result of heating or boiling the water over a period of time so that the water is evaporated, and the calcium and magnesium salts become concentrated and precipitate in the form of a "scale" on the walls of the container, hence the term "hardness". Water from different sources can have very different hardness values. Water samples can vary from simply being hard to being extremely hard. While this description of hardness is anything but quantitative, a quantitative description based on an EDTA titration can be given.

The EDTA determination of water hardness results from the reaction of the EDTA ligand with all of the metal ions involved: calcium, magnesium, and iron. An interesting question is: "How are the results reported"? Hardness is not usually reported precisely as so much calcium plus so much magnesium, *etc*. There is no distinction made between the metals involved. All species reacting with the EDTA are considered one species and the results reported as an amount of one species—calcium carbonate, $CaCO_3$. That is, in the calculation, when a molecular weight is used to convert mole to gram, the molecular weight of calcium carbonate is used, and thus a quantity of $CaCO_3$ equivalent to the sum of all contributors to the hardness is what is reported.

Ethylenediaminetetraacetic acid(EDTA)   EDTA comples with Ca^{2+} ($CaEDTA^{2-}$)

The indicator that is most often used is called eriochrome black T (EBT). EBT is actually a ligand that also reacts with the metal ions, like EDTA. In the free uncombined form, it imparts a sky blue color to the solution, but if it is a part of a complex ion with either calcium or magnesium ions, it is a wine red color. Thus, before adding any EDTA from a burette, the hard water sample containing the pH=10 ammonia buffer and several drops of EBT indicator will be wine red. As the EDTA solution is added, the EDTA ligand reacts with the free metal ions and then actually reacts with the metal-EBT complex ion, complexing the metal ions and resulting in the free EBT ligand, which, as mentioned earlier, gives a sky blue color to the solution. The color change, then, is the total conversion of the wine red color to the sky blue color, with every trace of red disappearing at the end point.

It is known that this color change is quite sharp when magnesium ions are present. In cases in which magnesium ions are not present in the water samples, the end point will not be sharp. Because of this, a small amount of magnesium chloride is added to the EDTA as it is prepared and thus a sharp end point is assured.

New Words

routine [ruːˈtiːn] *adj*. 常规的，平常的
dissolve [dɪˈzɒlv] *vt*. 解散，分散，（使）溶解
occasionally [əˈkeɪʒnəlɪ] *adv*. 偶然，偶尔
evaporate [ɪˈvæpəreɪt] *vt*. （使）蒸发，挥发
scale [skeɪl] *n*. 等级，刻度，标度
hence [hens] *adv*. 因此，之后
ligand [ˈlɪgənd] *n*. 配位体
distinction [dɪˈstɪŋkʃn] *n*. 差别，区分
equivalent [ɪˈkwɪvələnt] *adj*. 等同的，等效的
contributor [kənˈtrɪbjətə(r)] *n*. 作出贡献者，促成因素
impart [ɪmˈpɑːt] *v*. 传授，给予

Expressions and Technical Terms

water hardness 水的硬度
a significant quantity of... 大量的……
associate with... 与……联系在一起
hardness value 硬度值
quantitative description 定量描述
calcium carbonate 碳酸钙
molecular weight 分子量
eriochrome black T 铬黑T
ammonia buffer 氨缓冲溶液
complex ion 配离子

Exercises

A. Translate the following into Chinese or English.

1. EDTA titration   2. dissolved mineral   3. magnesium carbonate

4. quantitative description   5. calcium carbonate   6. ammonia buffer

7. 水硬度   8. 硬度值   9. 敏锐的终点   10. 配离子

B. Choose the best answer to each question.

1. What are routinely used to determine water hardness in a laboratory? (   )
   (A) Acid-base titrations   (B) EDTA titrations
   (C) Precipitation titrations   (D) Redox titrations

2. The problems that arise are mostly a result of heating or boiling the water over a period of time so that the water is (   ).
   (A) evaporated   (B) hot   (C) cool

3. 'Simply being hard' or 'extremely hard', this description of hardness is anything but (   ).
   (A) quantitative   (B) qualitative

4. All species reacting with the EDTA are considered one species and the results reported as an amount of one species. What is it? (   )
   (A) magnesium carbonate   (B) calcium carbonate   (C) iron

5. The color change is the total conversion of (   ) color to the sky blue color, with every trace of red disappearing at the end point.
   (A) the yellow   (B) water clear   (C) the wine red

C. Comprehension exercises.

1. How does the term "hardness" come out?

2. How are the results reported in the EDTA determination of water hardness?

Reading Material 阅读材料

Hard Water

Water containing dissolved calcium or magnesium ions is said to be hard. In hard water, soap does not lather as well and forms a scum. The calcium and magnesium ions get into the water in two ways. Firstly, rain falls on soluble calcium or magnesium minerals and dissolves them. Secondly, rain (which is slightly acidic) reacts with basic calcium or magnesium minerals. Rain water is acidic because it contains dissolved carbon dioxide. So rain is dilute carbonic acid.

The advantages of properly hard water are better taste, better for brewing beer and healthier (less heart disease, stronger bones and teeth). But its disadvantages are that it can form a scum with soap and forms a scale in hot water boilers and kettles.

Water which has had its hardness removed is described as softened water. Water is softened in two ways. Firstly, by adding sodium carbonate the calcium and magnesium ions are precipitated out. Secondly, the hard water is run through a resin which replaces the calcium or magnesium ions with sodium ions by using ion exchangers.

New Words

hard [hɑːd] *adj.* 困难的，硬的，有力的，努力的
dissolve [dɪˈzɒlv] *v.* 溶解，使溶解
calcium [ˈkælsɪəm] *n.* 钙
magnesium [mæɡˈniːzɪəm] *n.* 镁
ion [ˈaɪən] *n.* 离子
soap [səʊp] *n.* 肥皂
lather [ˈlɑːðə(r)] *v.* 起泡
scum [skʌm] *n.* 浮渣，泡沫，糟粕
mineral [ˈmɪnərəl] *n.* 矿物，矿石，矿物质，汽水
dilute [daɪˈluːt] *adj.* 稀释的，冲淡的
brew [bruː] *v.* 酿造
scale [skeɪl] *n.* 规模，比例（尺），级别，刻度数值范围，水垢
resin [ˈrezɪn] *n.* 树脂，合成树脂，松香

Expressions and Technical Terms

carbon dioxide 二氧化碳
heart disease 心脏疾病
be described as... 被描述为……
sodium carbonate 碳酸钠
precipitate out 沉淀出来
ion exchanger 离子交换剂

Exercises

Translate the following into English.

1. 硬水　　　2. 钙离子　　　3. 镁离子　　　4. 溶解

5. 水垢　　　6. 碳酸钠　　　7. 沉淀出来　　8. 离子交换法

Lesson 38

Analysis of Trace Amounts of Organic Materials in Soil and Sediments
土壤与沉积物中微量有机物的分析

Warming up

土壤是地球表层的岩石经过生物圈、大气圈和水圈的综合影响演变而成的，是指陆地表面具有肥力、能够生长植物的疏松表层，其厚度一般在 2m 左右。土壤能同时并不断地供应和调节植物在生长过程中所需要的水分、养分、空气、热量。其物质组成很复杂：由矿物质、有机物、水分、空气（固、液、气三相）组成。土壤是农业生产最基本的自然资源，能分解、转化污染物。

腐殖质：具有多种官能团、芳香族结构及酸性的高分子化合物。能改善土壤的物理、化学和生物学性状，具有吸附、缓冲和配合性能。

土壤污染具有隐蔽性和滞后性，难治理。土壤污染物有下列 4 类：①化学污染物，包括无机污染物和有机污染物；②物理污染物；③生物污染物；④放射性污染物。

土壤质量标准规定了土壤中污染物的最高允许浓度或范围，是判断土壤质量的依据。

测定方法：土壤水分——重量法；含量较高的成分——滴定法；重金属——分光光度法；有机氯、有机磷、有机汞等农药——气相色谱法。

环境作为一个整体，污染物进入环境，会在大气、水和土壤等部分进行迁移和转化运动，从而影响整个环境。因此，土壤监测必须与大气、水体、生物监测相结合，全面、客观地反映实际。

《土壤污染防治行动计划》是我国为了切实加强土壤污染防治，逐步改善土壤环境质量而制定的法规。

Text

From a biological viewpoint, trace elements are most conveniently classified into three groups: essential, non-essential, and toxic.

One of the major ever-increasing problems is the influence of chemical pollution, especially heavy metal contaminants, upon the complex interactions of the biosphere. The relationship between animal or human health and exposure to elements through air, water and food is an important area of environmental analytical research before assessing the requirements for trace element analysis.

While there are no specific guidelines for the analysis of trace organics in soil, any of the common methods of sample preparation can be used including headspace gas chromatography (GC) analysis and solvent extraction. Pesticides and polychlorinated biphenyls (PCBs) can be extracted with isooctane, while semi-volatiles like aromatic hydrocarbons are extractable with dichloromethane. Methods used for the analysis of volatile organic materials in water can be applied to the analysis of volatiles in wet sediments and soils.

Analysis of trace organic compounds in soil samples is greatly complicated by the sample matrix, and sample clean-up is very much dictated by the water and humic content of the soil. The analysis of semi- and non-volatile organics in dry or wet soil is relatively easier. Partition of analytes is often achieved *via* Soxhlet extraction with a non-polar solvent such as hexane. While analytical recoveries are common, the main problem with this technique is that high proportions of co-extractives are recovered, especially if the sample has a high humus content.

In general, different soil types, their physical and chemical properties, pH, water content, the plant species and soil organisms associated with the soil will influence the degree of trace element contamination, the distribution within the soil.

New Words

sediment [ˈsedɪmənt] *n.* 沉渣，（最终形成岩石层的沙石等）沉积物
essential [ɪˈsenʃl] *adj.* 必不可少的
toxic [ˈtɒksɪk] *adj.* 有毒的
contaminant [kənˈtæmɪnənt] *n.* 污染物，致污物
biosphere [baɪəʊsfɪə(r)] *n.* 生物圈
exposure [ɪkˈspəʊʒə(r)] *n.* 暴露，接触
assess [əˈses] *v.* 评价，评定
headspace [ˈhedspeɪs] *n.* 顶空（法）
pesticide [ˈpestɪsaɪd] *n.* 杀虫剂，农药
isooctane [ˌaɪsəʊˈɒkteɪn] *n.* 异辛烷
semi-volatile [ˌsemiˈvɒlətaɪl] *adj.* 半挥发性的
dichloromethane [daɪklɔːrəˈmeθeɪn] *n.* 二氯甲烷
volatile [ˈvɒlətaɪl] *adj.* （液体或固体）易挥发的
matrix [ˈmeɪtrɪks] *n.* 矩阵，模型
dictate [dɪkˈteɪt] *v.* 命令，影响，支配
humic [ˈhjuːmɪk] *adj.* 腐殖的
partition [pɑːˈtɪʃ(ə)n] *n.* 分离层

hexane [ˈheksein] n. 己烷
recovery [rɪˈkʌvəri] n. 回收利用（能源、化工品等）
humus [ˈhjuːməs] n. 腐殖质，腐殖土

Expressions and Technical Terms

trace amount of... 微量的……
organic material 有机物质，有机物料
heavy metal contaminant 重金属污染物
trace element analysis 微量元素分析
specific guideline 具体的指导
solvent extraction 溶剂萃取，溶剂抽提
polychlorinated biphenyl 多氯联苯
aromatic hydrocarbon 芳香烃
Soxhlet extraction 索氏提取
non-polar solvent 非极性溶剂
high proportion of... 高比例的……
soil organism 土壤生物

Exercises

A. Translate the following into Chinese or English.

1. 微量元素分析　　2. 有机物质　　3. 重金属　　4. 溶剂萃取

5. 非极性溶剂　　6. polychlorinated biphenyl　　7. Soxhlet extraction

8. aromatic hydrocarbon　　9. humic content　　10. soil organism

B. Decide whether the following statements are true (T) or false (F) according to the text. Write T for true and F for false in each blank.

(　　) 1. From a biological viewpoint, trace elements are most conveniently classified into two groups: essential, non-essential.

(　　) 2. The relationship between animal or human health and exposure to elements is from air.

(　　) 3. There are no specific guidelines for the analysis of trace organics in soil.

(　　) 4. Methods used for the analysis of volatile organic materials in water can be applied to the analysis of volatiles in wet sediments and soils.

(　　) 5. Hexane is a non-polar solvent.

C. Cloze.

1. Any of the common methods of sample preparation can be used including _____ GC

analysis and _____ extraction.

2. Partition of analytes is often achieved *via* Soxhlet extraction with a _____ solvent such as hexane.

3. Different soil types, their physical and chemical _____, pH, water content, the plant species and _____ organisms associated with the soil will influence the _____ of trace element contamination, the _____ within the soil.

Lesson 39
Determination of Protein in Foodstuffs by the Kjeldahl Method
食品中蛋白质的测定（凯氏定氮法）

Warming up

蛋白质（protein）是生命的物质基础，没有蛋白质就没有生命。因此，它是与生命、与各种形式的生命活动紧密联系在一起的物质。机体中的每一个细胞和所有重要组成部分都有蛋白质参与。蛋白质是生物体内重要的活性分子，包括催化新陈代谢的酵素和酶。

氨基酸（aminoacid）是构成蛋白质的基本单位，赋予蛋白质特定的分子结构形态，使其分子具有生化活性。

GB 5009.5—2016《食品安全国家标准 食品中蛋白质的测定》规定了食品中蛋白质的测定方法。标准第一法（凯氏定氮法）和第二法（分光光度法）适用于各种食品中蛋白质的测定，第三法（燃烧法）适用于蛋白质含量在 10g/100g 以上的粮食、豆奶粉、米粉、蛋白质粉等固体试样的测定。标准不适用于添加无机含氮物质、有机非蛋白质含氮物质的食品的测定。

凯氏定氮法

原理：食品中的蛋白质在催化加热条件下被分解，产生的氨与硫酸结合生成硫酸铵。碱化蒸馏使氨游离，用硼酸吸收后以硫酸或盐酸标准滴定溶液滴定，根据酸的消耗量计算氮含量，再乘以换算系数，即为蛋白质的含量。

Text

The basic components of food include carbohydrates, proteins, lipids, vitamins, minerals and water needed for human nutrition, providing substances and energy needed for normal metabolism.

Proteins are the physical building blocks of life. Protein is the material basis of life, an important component of biological somatic tissue, and the raw material for organism development and tissue repair. They can regulate the acid and alkali in the human body and the moisture level, transfer genetic information and promote the human body material metabolism and transport.

At present, the main method of food testing includes sensory detection method, physical method, chemical method and instrumental detection methods and other different types of methods.

National Food Safety Standard (GB 5009. 5—2016) adopts 'Kjeldahl Method' as the first method to determine the content of protein in foodstuffs.

Principle: Proteins in food are broken down under catalytic heating conditions, and the resulting ammonia combines with sulfuric acid to form ammonium sulfate. Alkalization distills free ammonia. After absorption with boric acid, it is titrated with sulfuric acid or hydrochloric acid standard solution. Calculate the content of nitrogen according to the acid consumption, multiplying by conversion coefficient, and convert to the content of protein.

Instruments and equipment: Automatic Kjeldahl apparatus, balance.

Method: The well mixed solid sample was 0.2-2g, or the semisolid sample was 2-5g, or the liquid sample was 10-25g (about 30-40mg nitrogen), all accurate to 0.001g. Put one of them into the digestive tube, and then add 0.4g copper sulfate, 6g potassium sulfate and 20mL sulfuric acid in the digestion furnace. When the furnace temperature reaches 420℃, continue digestion for 1h. At this time, the liquid in the digestive tube is green and transparent. Take it out, after cooling, and add 50mL water to automatic Kjeldahl apparatus that is fed with sodium hydroxide solution, hydrochloric acid or sulfuric acid standard solution and boric acid solution containing mixed indicator before use. Then, automatic Kjeldahl apparatus achieve automatic feeding, distillation, titration and recording titration data.

Calculation: The determination result is the total nitrogen. The total nitrogen is multiplied by the protein conversion coefficient to get the protein content.

Kjeldahl method is suitable for the determination of protein content in all kinds of food, not suitable for the determination of food samples with inorganic nitrogen substances and organic non-protein nitrogen substances. Treating the sample with a solution of trichloroacetic acid allows the actual proteins to form a precipitate.

New Words

kjeldahl [kdʒel'deɪ] *n.* 凯氏
protein ['prəʊtiːn] *n.* 蛋白质
carbohydrate [ˌkɑːbəʊ'haɪdreɪt] *n.* 碳水化合物，糖类
lipid ['lɪpɪd] *n.* 脂质，油脂
nutrition [njuːˈtrɪʃ(ə)n] *n.* 营养
metabolism [məˈtæbəlɪzəm] *n.* 新陈代谢
regulate ['regjʊleɪt] *vt.* 管理，调节
adopt [əˈdɒpt] *vt.* 收养，采取，采纳
alkalization [ˌælkəlaɪˈzeɪʃən] *n.* 碱性化
multiply ['mʌltɪplaɪ] *vt.* 乘
coefficient [ˌkəʊɪˈfɪʃ(ə)nt] *n.* 系数
semisolid [ˌsemɪˈsɒlɪd] *adj.* 半固体的
trichloroacetic [traɪˌklɔːrəʊəˈsiːtɪk] *n.* 三氯乙酰

Expressions and Technical Terms

Kjeldahl method 凯氏定氮法
biological somatic tissue 生物体细胞组织
organism development 生物体发育
tissue repair 组织修复
moisture level 水分含量
genetic information 遗传信息，基因信息
sensory detection method 感官检测方法
boric acid 硼酸
conversion coefficient 转换系数
automatic Kjeldahl apparatus 自动凯氏定氮仪
digestive tube 消化管
digestion furnace 消化炉
titration data 滴定数据
trichloroacetic acid 三氯乙酸

Exercises

A. Translate the following into English or Chinese.

1. 蛋白质   2. 碳水化合物   3. 营养   4. 感官检测法

5. 消化管   6. 滴定数据   7. 蒸馏   8. 转换系数

9. Kjeldahl method   10. tissue repair   11. genetic information

12. ammonium sulfate   13. boric acid   14. hydrochloric acid

15. potassium sulfate   16. automatic Kjeldahl apparatus

B. Decide whether the following statements are true (T) or false (F) according to the text. Write T for true and F for false in each blank.

(　　) 1. Proteins are the physical building blocks of life and the material basis of life.

(　　) 2. Calculate the content of nitrogen according to the acid consumption and it is the content of protein.

(　　) 3. The main instruments and equipment are automatic Kjeldahl apparatus and a balance.

(　　) 4. The solid sample should be well mixed.

(　　) 5. Kjeldahl method is suitable for the determination of all kinds of nitrogen substances.

C. Translation.

1. The basic components of food include carbohydrates, proteins, lipids, vitamins, minerals and water.

2. At present, the main method of food testing includes sensory detection method, physical method, chemical method and instrumental detection methods and other different types of methods.

3. National Food Safety Standard adopts "Kjeldahl method" as the first method to determine the content of protein in foodstuffs.

译　文

凯氏定氮法测定食品中的蛋白质

食物的基本成分包括人体营养所需的碳水化合物、蛋白质、脂类、维生素、矿物质和水，提供人体正常新陈代谢所需的物质和能量。

蛋白质是生命的物质基础，是生物体细胞组织的重要组成部分，是机体发育和组织修复的原料。它们能调节人体内的酸碱和水分水平，传递遗传信息，促进人体物质的代谢和运输。

目前，食品检测的主要方法有感官检测法、物理检测法、化学检测法和仪器检测法等不同类型的方法。

GB 5009.5—2016《食品安全国家标准 食品中蛋白质的测定》采用"凯氏定氮法"作为测定食品中蛋白质含量的第一法。

原理：食物中的蛋白质在催化加热条件下被分解，产生的氨与硫酸结合形成硫酸铵。碱化蒸馏使氨游离，硼酸吸收后，用硫酸或盐酸标准滴定溶液滴定。根据酸的消耗量计算出氮的含量，乘以蛋白质转换系数，转换成蛋白质的含量。

仪器设备：自动凯氏定氮仪、天平

方法：称取充分混匀的固体试样 0.2～2g、半固体试样 2～5g 或液体试样 10～25g（约等于 30～40mg 氮），精确至 0.001g，至消化管中，再加入 0.4g 硫酸铜、6g 硫酸钾及 20mL 硫酸于消化炉进行消解。当消化炉温度达到 420℃之后，继续消解 1h，此时消化管中的液体呈绿色透明状，取出冷却后加入 50mL 水，于自动凯氏定氮仪（使用前加入氢氧化钠溶液，盐酸或硫酸标准溶液以及含有混合指示剂的硼酸溶液）上实现自动加液、蒸馏、滴定和记录滴定数据的过程。

计算：测定结果为总氮。总氮乘以蛋白质转换系数得到蛋白质含量。

凯氏定氮法适用于各种食品中蛋白质含量的测定，不适用于添加无机含氮物质、有机非蛋白质含氮物质的食品的测定。用三氯乙酸溶液处理样品可以使实际的蛋白质形成沉淀。

Lesson 40

Determination of Soluble Sugar in Cereals and Beans Using Iodometry
谷物和豆类中可溶性糖的测定（碘量法）

Warming up

糖是食品中重要的营养成分，可溶性糖（soluble sugar）是指在生物细胞内呈溶解状态，可被水和其他极性溶剂提取出来的糖，单糖和寡糖均为可溶性糖。

可溶性糖是谷物、豆类中重要的营养成分之一，其对人体的生理作用不容忽视。尽管可溶性糖在谷物、豆类中的含量并不高，但其对植物生长、品质鉴定、加工生产及遗传育种方面都起到了重要的作用。

GB/T 37493—2019《粮油检验 谷物、豆类中可溶性糖的测定 铜还原-碘量法》中规定了谷物、豆类中可溶性糖的测定方法：间接碘量法。

碘量法： 以碘作为氧化剂，或以碘化物（如碘化钾）作为还原剂进行滴定的方法，用于测定物质含量，是一种氧化还原滴定法。碘量法可用于测定水中游离氯、总氯、溶解氧，气体中硫化氢，食品中维生素C、葡萄糖等物质的含量。因此碘量法是环境、食品、医药、冶金、化工等领域最为常用的监测方法之一。

碘量法分为直接碘量法和间接碘量法，其中间接碘量法又分为剩余碘量法和置换碘量法。

直接碘量法： 是用碘滴定液直接滴定还原性物质的方法。在滴定过程中，I_2 被还原为 I^-：

$$I_2 + 2e^- \rightleftharpoons 2I^-$$

间接碘量法

（1）剩余碘量法：是一定条件下在待测样品（还原性物质）溶液中先加入定量、过量的碘滴定液，待 I_2 与待测组分反应完全后，用硫代硫酸钠标准溶液滴定剩余的碘，以求出待测组分含量的方法。

（2）置换碘量法：是一定条件下先在待测样品（氧化性物质）溶液中加入碘化钾，待测组分将碘化钾氧化析出定量的碘，碘再用硫代硫酸钠标准溶液滴定，从而可求出待测组分含量。

指示剂： 淀粉。淀粉指示剂应在近终点时加入，因为当溶液中有大量碘存在时，碘易吸附在淀粉表面，影响终点的正确判断。极微量的碘与多羟基化合物淀粉相遇，能立即形成深蓝色的配合物，这一性质在碘量法中得到应用。

Text

Soluble sugar is one of the important nutrition in cereals and beans. The physiological effects on human body can not be ignored, and it plays an important role in plant growth, quality identification, processing, production, genetics and breeding.

Grain and oils inspection standard (GB/T 37493—2019) adopts iodometry (an indirect method) to determine the content of soluble sugar in cereals and beans.

Principle: Soluble sugar in cereals and beans is hydrolyzed into reducing sugar, and it reacts with alkaline copper reagent (Cu^{2+}) to generate cuprous oxide (Cu_2O) precipitation. Under the condition of acidic sulfate, cuprous oxide reacts with potassium iodate (KIO_3) and potassium iodide (KI) to form iodine (I_2) quantitatively. The rest of the iodine is titrated with sodium thiosulfate standardized solution. Replace the test sample with blank water for the same operation as blank experiment.

An important reactant in redox titrimetry is potassium iodide, KI. KI is a reducing agent that is useful in analyzing for oxidizing agents. The interesting aspect of the iodide is that it is most often used in an indirect method. This means that the oxidizing agent is not measured directly by a titration with KI, but is measured indirectly by the titration of the iodide that forms in the reaction. The KI is actually added in excess, since it need not be measured at all. The experiment is called iodometry. Thus the percent of the oxidizing agent is calculated indirectly from the amount of titrant since the titrant actually reacts with I_2 and not "oxidizing agent". This titrant is normally sodium thiosulfate ($Na_2S_2O_3$).

The sodium thiosulfate ($Na_2S_2O_3$) solution must be standardized. Several primary standard oxidizing agents are useful for this. Probably the most common one is potassium dichromate ($K_2Cr_2O_7$). Primary standard potassium bromate $KBrO_3$, or potassium iodate KIO_3, can also be used. Even primary standard iodine, I_2, can also be used. Usually in the standardization procedures, KI is again added to the substance to be titrated and the liberated iodine titrated with thiosulfate. If I_2 is the primary standard, it is titrated directly. The end point is usually detected with the use of a starch solution as the indicator. Starch, in the presence of iodine, is a deep blue color. When all iodine consumed by the thiosulfate, the color changes sharply from blue to violet providing a very satisfactory end point. Some important precautions concerning the starch, however, are to be considered. The starch solution should be fresh, should not be added until the end point is near, cannot be used in strong acid solutions, and cannot be used with solution temperatures above 40℃.

$$KI + "O" \longrightarrow "R" + I_2$$
$$I_2 + 2Na_2S_2O_3 \longrightarrow 2NaI + Na_2S_4O_6$$

An iodometry experiment: a solution of KI is added to a solution of the substance to be determined (represented by "O"). The reaction products are "R" and iodine, I_2. The amount of I_2, which is proportional to the amount of "O", is then titrated with thiosulfate, $Na_2S_2O_3$.

New Words

cereal ['sɪərɪəl] *n.* 谷类食物

iodometry [ˌaɪə'dɒmətrɪ] n. 碘量法
nutrition [njʊ'trɪʃ(ə)n] n. 营养，滋养
physiological [ˌfɪzɪə'lɒdʒɪk(ə)l] adj. 生理的，生理机能的
ignore [ɪg'nɔː(r)] v. 佯装未见，忽视
breed [briːd] v. 繁殖
hydrolyze ['haɪdrəlaɪz] vt. 使水解
iodine ['aɪədiːn] n. 碘
titrimetry [taɪ'trɪmɪtrɪ] n. 滴定测量
direct [də'rekt] adj. 直接的
titrate [taɪ'treɪt] v. 用滴定法测量
standardize ['stændədaɪz] vt. 使标准化，标定
liberate ['lɪbəreɪt] vt. 解放，释放，释出，放出
detect [dɪ'tekt] vt. 查明，探测，洞察，侦察
indicator ['ɪndɪkeɪtə(r)] n. 指示剂
consume [kən'sjuːm] vt. 消耗，耗尽
precaution [prɪ'kɔːʃn] n. 预防，警惕，预防措施
amount [ə'maʊnt] n. 总数，量

Expressions and Technical Terms

soluble sugar 可溶性糖
quality identification 品质鉴定
grain and oil inspection standard 粮油检验标准
reducing sugar 还原糖
cuprous oxide 氧化亚铜
potassium iodate 碘酸钾
potassium iodide 碘化钾
sodium thiosulfate 硫代硫酸钠
standardized solution 标准溶液
blank experiment 空白实验
indirect method 间接法
reducing agent 还原剂
oxidizing agent 氧化剂
in excess 过量地
primary standard 基准物
potassium dichromate 重铬酸钾
potassium bromate 溴酸钾
end point 滴定终点
starch solution 淀粉溶液
in the presence of... 在……存在时
change sharply 突变

strong acid solution 强酸溶液
be proportional to 与……成比例的

✤ Exercises

A. Translate the following into English.

1. 可溶性糖　　　　2. 品质鉴定　　　　3. 还原糖　　　　4. 氧化剂

5. 还原剂　　　　　6. 氧化还原滴定　　7. 间接法　　　　8. 滴定终点

9. 过量地　　　　　10. 硫代硫酸钠　　　11. 标准溶液　　　12. 淀粉溶液

13. 突变　　　　　　14. 碘酸钾　　　　　15. 溴酸钾　　　　16. 碘化钾

B. Choose the best answer to each question.
1. In redox titrimetry, KI is a (　　) that is useful in analyzing for oxidizing agents.
　(A) reducing agent　　　(B) oxidizing agent
2. In iodometry, the titrant is normally (　　).
　(A) sodium hydroxide　　(B) sodium thiosulfate　　(C) silver ions
3. The sodium thiosulfate ($Na_2S_2O_3$) solution must be standardized. Probably the most common primary standard oxidizing agent is (　　).
　(A) sodium carbonate　　(B) zinc oxide　　(C) potassium dichromate
4. In iodometry, the end point is usually detected with the use of a (　　) as the indicator.
　(A) starch solution　　　(B) eriochrome black T　　(C) methyl orange
5. When all iodine consumed by the thiosulfate, the color changes sharply from (　　) to (　　) providing a very satisfactory end point.
　(A) blue　　　　　　　　(B) violet　　　　　　　　(C) red

C. Translate following sentences into Chinese.
1. Replace the test sample with blank water for the same operation as blank experiment.

2. Indirect method means that the oxidizing agent is not measured directly by a titration with KI, but is measured indirectly by the titration of the iodide that forms in the reaction.

3. The starch solution should be fresh, should not be added until the end point is near, cannot be used in strong acid solutions, and cannot be used with solution temperatures above 40℃.

译 文

谷物、豆类中可溶性糖的测定——碘量法

可溶性糖是谷物、豆类中重要的营养成分之一，其对人体的生理作用不容忽视，它对植物生长、品质鉴定、加工生产及遗传育种方面都起到了重要的作用。

GB/T 37493—2019《粮油检验 谷物、豆类中可溶性糖的测定 铜还原-碘量法》中规定用碘量法（间接碘量法）测定谷物、豆类中可溶性糖的含量。

谷物、豆类中的可溶性糖被水解成还原糖后与碱性铜试剂中的 Cu^{2+} 作用，生成氧化亚铜（Cu_2O）沉淀。在硫酸的酸性条件下，氧化亚铜能定量地消耗碘酸钾和碘化钾生成的碘，溶液中剩余的碘用硫代硫酸钠标准溶液滴定，同时以水代替样液做空白滴定。

氧化还原滴定法中，一种重要的反应物是碘化钾。碘化钾在分析氧化剂时是一种有用的还原剂。碘化物有趣的一面是它最常被用在间接方法中。这意味着被分析的氧化剂不直接用碘化钾滴定来测定，而是通过滴定反应中形成的碘来间接测定。由于碘化钾不需被测量，因而实际上被过量加入。这一实验被称为碘量滴定法。由于滴定剂实际上是和碘而不是和氧化剂反应，因而，根据滴定剂的量，氧化剂的比例可被间接计算出。这种滴定剂通常是 $Na_2S_2O_3$。

硫代硫酸钠必须被标定。几种基准物的氧化剂被用于这一过程中。最常用的是重铬酸钾。基准物溴酸钾或碘酸钾也能用于标定，甚至基准物的碘也能用于标定。在标定程序中，碘化钾加入被滴定的物质中，释放出的碘用硫代硫酸钠来滴定。如果碘是基准物，可直接被滴定。终点通常用指示剂淀粉溶液来检测。淀粉遇到碘呈深蓝色。当所有的碘被硫代硫酸盐消耗掉时，颜色明显地由蓝变成紫色，提供了一个很明显的终点。然而，有关淀粉需要采取一些重要的预防措施。淀粉溶液应是新鲜的，应到接近终点时才加入，不能用于强酸溶液中，不能用在40℃以上的溶液中。

$$KI + \text{"O"} \longrightarrow \text{"R"} + I_2$$
$$I_2 + 2Na_2S_2O_3 \longrightarrow 2NaI + Na_2S_4O_6$$

碘量滴定法实验：碘化钾被加入待测定溶液（用"O"来表示）中。反应产物是碘和用"R"表示的还原物。碘的量与氧化剂的量成比例，可用硫代硫酸钠来滴定。

Lesson 41
Pharmacopoeia and Analysis of Aspirin
药典简介和阿司匹林的分析检验

Warming up

药典（pharmacopoeia）是一个国家记载药品标准、规格的法典，一般由国家药品监督管理局主持编纂、颁布实施，国际性药典则由公认的国际组织或有关国家协商编订。制定药品标准对加强药品质量的监督管理、保证质量、保障用药安全有效、维护人民健康起着十分重要的作用。药品标准一般包括以下内容：法定名称、来源、性状、鉴别、纯度检查、含量（效价或活性）测定、类别、剂量、规格、贮藏、制剂，等等。

1949 年中华人民共和国成立后，《中华人民共和国药典》（简称《中国药典》）已有十一个版次，目前中国药典为 2020 版。目前常用药典：国际药典（Ph. Int）、美国药典（USP）、英国药典（BP）和欧洲药典（Ph. Eup）等。

《中国药典》2020 年版由一部、二部、三部、四部构成，收载品种总计 5608 种，其中新增 1082 种。本版药典继续秉承保护野生资源和自然环境、坚持中药可持续发展、倡导绿色标准的理念；重点加强药品安全性和有效性的控制要求，充分借鉴国际先进的质量控制技术和经验，全面反映了我国当前医药发展和检测技术的现状。

阿司匹林（$C_9H_8O_4$）〔aspirin，2-（乙酰氧基）苯甲酸，又名乙酰水杨酸〕是一种白色结晶或结晶性粉末，无臭或微带醋酸臭，微溶于水，易溶于乙醇，可溶于乙醚、氯仿，水溶液呈酸性。阿司匹林已应用百年，成为医药史上三大经典药物之一，至今它仍是世界上应用最广泛的解热、镇痛和抗炎药，也是作为比较和评价其他药物的标准制剂。

乙酰水杨酸(阿司匹林)

Text

Pharmacopoeia is a book containing directions for the identification of compound medicines, and published by the authority of a government or a medical or pharmaceutical society, such as *Chinese Pharmacopoeia*, *United States Pharmacopoeia* (USP) and *British Pharmacopoeia* (BP).

Part 4 Analysis and Inspection Technology 分析检验技术

Drug Inspection Procedures are sampling, inspection, sample retention and report.

Sampling: Reasonable sampling according to the total number of samples requires scientificity, truthfulness and representation.

Inspection: Test samples according to drug standards, are firstly checked whether the traits meet the requirements, and then identified, checked, and determined the content.

Sample retention: Samples must be retained for received inspections, and the number of samples to be retained must not be less than the amount of full inspection at one time.

Report: True, complete, concise, specific, format specification.

Determination of Aspirin

1. Acid-base titration

Principle: The free carboxyl group of aspirin is acidic and can form a salt with a base. It can be titrated directly with sodium hydroxide titrant. Aspirin is an organic acid. When titrated with sodium hydroxide, the stoichiometric point is a little alkaline, so phenolphthalein, which changes color in the alkaline zone is used as the indicator.

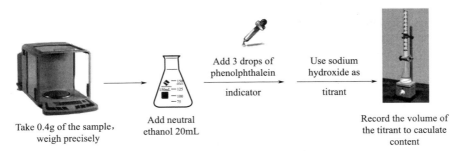

Method: Weigh accurately about 0.4g and dissolve in 20mL neutral ethanol, add 3 drops of phenolphthalein indicator and titrate with 0.1mol/L sodium hydroxide titrant. Each milliliter of 0.1mol/L sodium hydroxide titrant is equivalent to 18.02mg of $C_9H_8O_4$.

Calculation: Percentage content of aspirin.

2. High performance liquid chromatography *Chinese Pharmacopoeia* 2020 edition adopts high performance liquid chromatography to determine the content of aspirin tablets and aspirin enteric-coated tablets.

Principle: A small amount of tartaric acid or citric acid (as a stabilizer) added to aspirin tablets, as well as salicylic acid and acetic acid that may be produced in the preparation process can consume the alkali titrant, which makes the determination result high.

Column: C_{18}.

Mobile phase: Acetonitrile-tetrahydrofuran-acetic acid-water (20:5:5:70).

Test wavelength: 276nm.

Solution preparation: Take 20 tablets of this product, accurately weigh it, grind it thoroughly, accurately weigh an appropriate amount of fine powder (about 10mg of aspirin), and place it in a 100mL measuring flask. Shake it strongly with 1% glacial acetic acid methanol solution to dissolve aspirin, and add the 1% glacial acetic acid methanol solution to the mark. Shake well, filter through the membrane, and take the additional filtrate as the test solution.

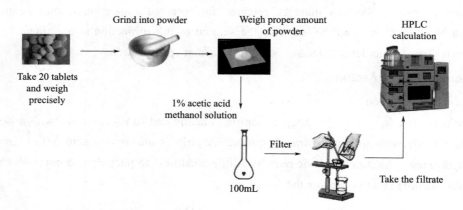

Calculation: external standard method.

New Words

pharmacopoeia [ˌfɑːməkəˈpiːə] n. 药典，处方书
pharmaceutical [ˌfɑːməˈsuːtɪkl] adj. 制药的
retention [rɪˈtenʃ(ə)n] n. 保留，保存
scientificity [ˌsaɪəntɪˈfɪsətɪ] n. 科学性
trait [treɪt] n. 特征，特点
retain [rɪˈteɪn] v. 保留，储存
concise [kənˈsaɪs] adj. 简明的，简洁的
specific [spəˈsɪfɪk] adj. 明确的，具体的
phenolphthalein [ˌfiːnɒlˈ(f)θæliːn] n. 酚酞
tablet [ˈtæblət] n. 药片，片剂
enteric-coated [enˈterɪk ˈkəʊtɪd] adj. （包有）肠溶（衣）的
tartaric [tɑːˈtærɪk] adj. 酒石的
citric [ˈsɪtrɪk] adj. 柠檬的
stabilizer [ˈsteɪbəlaɪzə(r)] n. 稳定剂
salicylic [ˌsælɪˈsɪlɪk] adj. 水杨酸的
acetic [əˈsiːtɪk] adj. 乙酸的
acetonitrile [əˌsiːtəʊˈnaɪtrɪl] n. 乙腈
tetrahydrofuran [ˌtetrəˌhaɪdrəˈfjʊəræn] n. 四氢呋喃
grind [ɡraɪnd] v. 磨碎
ground [ɡraʊnd] v. 研磨（grind 的过去式和过去分词）
thorough [ˈθʌrə] adj. 彻底的，全面的

glacial ['gleɪʃ(ə)l] *adj.* 冰的，冰河时代的
filter ['fɪltə(r)] *v.* 过滤；*n.* 过滤器
filtrate ['fɪltreɪt] *n.* 滤液

Expressions and Technical Terms

compound medicine 复方药
authority of a government 政府的权威部门
pharmaceutical society 药学会
Chinese Pharmacopoeia 中华人民共和国药典
drug inspection procedure 药品检验程序
drug standard 药品标准
sample retention 留样
received inspection 收到检验（对象）
full inspection 全分析
format specification 格式规格
carboxyl group 羧基
an organic acid 一种有机酸
the stoichiometric point 化学计量点
alkaline zone 碱性范围
neutral ethanol 中性乙醇
high performance liquid chromatography 高效液相色谱法
measuring flask 容量瓶
external standard method 外标法

Exercises

A. Translate the following into Chinese or English.

1. *Chinese Pharmacopoeia*   2. enteric-coated   3. neutral ethanol

4. drug inspection procedure   5. external standard method

6. tartaric acid   7. citric acid   8. salicylic acid   9. acetic acid

10. 羧基   11. 留样   12. 全分析   13. 容量瓶

B. Choose the best answer to each question.

1. Drug inspection procedures are (   ).
(A) Sampling   (B) Inspection   (C) Sample retention   (D) Report

2. The number of samples to be retained must not be (   ) the amount of full inspection at one time.
(A) less than   (B) more than   (C) equal to

3. The free carboxyl group of aspirin is acidic and can form a salt with a (   ).
（A）acid　　　　　（B）base　　　　　（C）salt

4. *Chinese Pharmacopoeia* 2020 edition adopts (   ) to determine the content of aspirin tablets and aspirin enteric-coated tablets.
（A）high performance liquid chromatography　　（B）acid-base titration

C. Translation.

1. Reasonable sampling according to the total number of samples requires scientificity, truthfulness and representation.

2. A small amount of acid may be produced in the preparation process can consume the alkali titrant, which makes the determination result high.

3. Take 20 tablets of this product, accurately weigh it, grind it thoroughly, and accurately weigh an appropriate amount of fine powder（about 10mg of aspirin）.

Lesson 42
Introduction to Chinese Medicine Analysis and Inspection
走进中药的分析检验

Warming up

中药： 以中国传统医药理论指导采集、炮制、制剂，说明作用机理，指导临床应用的药物，统称为中药。中药就是指在中医理论指导下，用于预防、治疗、诊断疾病并具有康复与保健作用的物质。

中药主要来源于天然药及其加工品，包括植物药、动物药、矿物药及部分化学、生物制品类药物。由于中药以植物药居多，故有"诸药以草为本"的说法。

天然药材的分布和生产离不开一定的自然条件。中药的采收时节和方法对确保药物的质量有着密切的关联。把药物与疗效有关的性质和性能统称为药性。

中药分析与检测的基本程序： 取样、样品的制备、定性鉴定、检查和含量测定。

Text

Chinese medicine is a treasure of China's excellent traditional culture. Many ancient books of Chinese medicine describe a lot of traditional Chinese medicine and prescriptions, which have great guidance and research value for ancient and modern traditional Chinese medicine.

Chinese medicine is the substance with the function of rehabilitation and health care, applied for prevention, treatment and diagnosis of diseases under the guidance of Chinese medicine theory.

Chinese medicine analysis is the discipline and the important branch of pharmaceutical analysis to study Chinese medicinal materials and pieces, extracts and Chinese medicine preparations by modern analysis methods.

Compared with chemical drugs, traditional Chinese medicine has the following three characteristics: (1) Complex composition; (2) Dose effect correlation is not clearly elucidated;

(3) Many factors affect the active ingredients of medicinal herbs.

Impurities refer to substances that are introduced during the production or storage of the drug, have no therapeutic effect or affect the stability and efficacy of the drug, or are even harmful to human body. This is the same as the concept as impurities in chemical drugs.

Typical inspection items of traditional Chinese medicine: (1) Moisture test; (2) Ash test; (3) Pesticide residues; (4) Endogenous harmful substances.

Authenticity identification: (1) Characters and microscopic identification are unique methods of Chinese materia medica analysis; (2) Physicochemical medicine identification method: using physical chemical or physicochemical methods to identify authenticity; (3) Chromatographic identification method has the double advantages of separation and analysis. It is widely used in the identification of Chinese materia medica.

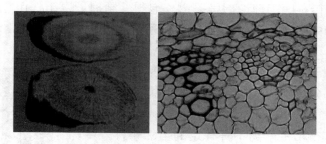

The task of Chinese medicine analysis is discriminating real and fake of Chinese herbs, controlling the quality of Chinese medicine, guiding production, research PK-PD and finding new medicine sources.

New Words

prescription [prɪˈskrɪpʃn] *n.* 处方，药方

rehabilitation [riːəˌbɪlɪˈteɪʃn] *n.* 康复，恢复，修复

prevention [prɪˈvenʃn] *n.* 预防，防止

diagnosis [ˌdaɪəgˈnəʊsɪs] *n.* 诊断，判断

discipline [ˈdɪsəplɪn] *n.* 纪律，（尤指大学的）科目，学科

pharmaceutical [ˌfɑːməˈsuːtɪkl] *adj.* 制药的

piece [piːs] *n.* 块，张，片，段

extract [ˈekstrækt] *n.* 提取物，汁

correlation [ˌkɒrəˈleɪʃ(ə)n] *n.* 相互关系，关联

elucidate [ɪˈluːsɪdeɪt] *v.* 阐明，解释

therapeutic [ˌθerəˈpjuːtɪk] *adj.* 治疗的，有疗效的

efficacy [ˈefɪkəsɪ] *n.* 功效，效力

inspection [ɪnˈspekʃ(ə)n] *n.* 检查

endogenous [enˈdɒdʒənəs] *adj.* 内生的，内因性的

unique [jʊˈniːk] *adj.* 独特的

authenticity [ˌɔːθenˈtɪsətɪ] *n.* 真实性，可靠性

discriminate [dɪˈskrɪmɪneɪt] *v.* 区分，辨别

Part 4 Analysis and Inspection Technology 分析检验技术

Expressions and Technical Terms

Chinese medicine 中药
China's excellent traditional culture 中华优秀传统文化
research value 研究价值
dose effect 剂量效应
medicinal herb 药草，草药
active ingredient 有效成分
refer to 提到，谈及
moisture test 水分测定
ash test 灰分测定
pesticide residue 杀虫剂残留
character and microscopic identification 性状和显微鉴别
materia medica 药物学
chromatographic identification method 色谱鉴别方法
PK-PD 药代动力学-药效动力学

Exercises

A. Translate the following into Chinese or English.

1. 中药
2. 剂效关系
3. 研究价值
4. 水分测定

5. 灰分测定
6. inspection items
7. authenticity identification

8. pesticide residues
9. characters and microscopic identification

B. Decide whether the following statements are true （T） or false （F）. Write T for true and F for false in each blank.

（   ） 1. Chinese medicine is a treasure of China's excellent traditional culture.

（   ） 2. Dose-effect correlation of traditional Chinese medicine is clearly elucidated.

（   ） 3. Many factors affect the active ingredients of medicinal herbs.

（   ） 4. Impurities refer to substances are usually introduced during the production or storage of the drug.

（   ） 5. Endogenous harmful substances is one of typical inspection items of traditional Chinese medicine.

C. Cloze.

1. Compared with chemical drugs，traditional Chinese medicine has the following three characteristics：（1） _____ composition；（2） _____ correlation is not clearly elucidated；（3） Many factors affect the active _____ of medicinal herbs.

2. Authenticity identification：（1） _____ and _____ identification are unique methods

of Chinese materia medica analysis; (2) _____ medicine identification method; (3) Chromatographic identification method has the double advantages of _____ and _____ .

D. Translation.

1. Typical inspection items of traditional Chinese medicine: (1) Moisture test; (2) Ash test; (3) Pesticide residues; (4) Endogenous harmful substances.

2. The task of Chinese medicine analysis is discriminating real and fake of Chinese herbs, controlling the quality of Chinese medicine, guiding production, research PK-PD and finding new medicine sources.

Lesson 43
Analysis of the Bisphenol A
双酚 A 的分析检验

Warming up

4,4-二羟基二苯基丙烷 [bisphenol A (BPA), 又称双酚 A, 是一种有机化合物, 分子式为 $C_{15}H_{16}O_2$] 是聚碳酸酯、环氧树脂等多种高分子材料的原料, 这些高分子材料被广泛用于生产化工产品和食品相关产品, 如食品包装材料及容器。 在塑料制品的制造过程中, 添加双酚 A 可以使其具有无色透明、耐用、轻巧和突出的防冲击性等特性, 尤其能防止酸性蔬菜和水果从内部侵蚀金属容器。 双酚 A 可通过食品包装材料及容器迁移至食品中, 食品相关产品国家标准规定了其迁移量。

我国卫生部等 6 部门关于禁止双酚 A 用于婴幼儿奶瓶的公告（公告 2011 年第 15 号）

新规定称, 拟自 2011 年 6 月 1 日起, 禁止双酚 A 用于婴幼儿食品容器（如奶瓶）生产和进口。 自 2011 年 9 月 1 日起, 禁止销售含双酚 A 的婴幼儿食品容器。 不过, 双酚 A 允许用于生产除婴幼儿奶瓶以外的其他食品包装材料、容器和涂料, 迁移量应当符合相关食品安全国家标准规定的限量。

科学研究表明, 食品相关产品中迁移的双酚 A 极其微量, 尚未发现双酚 A 对人体健康产生不良影响。 鉴于婴幼儿属于敏感人群, 为防范食品安全风险, 保护婴幼儿健康, 现决定禁止双酚 A 用于婴幼儿奶瓶。

双酚 A 在生活中应用广泛, 成为人们经常能接触到的物质。 因此, 其安全性问题成了公众的关注的焦点, 但其应用目前仍存在争议。

Text

Bisphenol A (BPA) is a monomeric intermediate in the production of polycarbonate plastics and epoxy resins. Epoxy resins are used to coat the inside of metal food cans, and polycarbonate plastics are often used to store food, such as plastic baby bottles, plastic water bottles, and clear plastic tableware.

The primary exposure of BPA to humans is through the diet, although there can be some exposure from air, water, and dust.

Animal studies have shown that BPA mildly mimics the female hormone estrogen. Test animals exposed to BPA have been shown to develop precancerous changes. Because of the safety concerns over BPA, BPA-free products have been introduced commercially. The two most common replacements are bisphenol S (BPS) and bisphenol F (BPF).

In all cases, reliable analytical methods are essential to determine BPA (and its replacements) in food and beverage containers, human body fluids, and the environment, including the water supply. Determination of BPA, BPS, and BPF is made difficult because of the variety of matrices in which these chemicals are found. Most of the methods for BPA use some form of extraction to eliminate much of the matrix and concentrate the analyte. Extraction is usually followed by a chromatographic separation step with fluorescence or mass spectrometry detection. The most sensitive and specific methods use GC and HPLC with MS or MS/MS detection. Although the published determinations of BPS and BPF are scarce, methods similar to those for BPA should be applicable since the compounds are so structurally similar.

The sensitivity and specificity of methods for BPA and its substitutes have improved greatly with the introduction of tandem mass spectrometric methods and ultrahigh-pressure LC. More research needs to be done on the human health hazards of BPA, BPS, and BPF and their related metabolites.

New Words

bisphenol [ˈbɪsfɪnɒl] n. 双酚
monomeric [ˌmɒnəˈmerɪk] adj. 单体的，单分子构造的
epoxy [ɪˈpɒksɪ] adj. 环氧的，环氧树脂的
tableware [ˈteɪblweə(r)] n. 餐具
mild [maɪld] adj. 轻微的
mimic [ˈmɪmɪk] vt. 模仿，模拟
estrogen [ˈiːstrədʒən] n. 雌性激素
precancerous [ˌpriːˈkænsərəs] adj. 癌症前期的
commercially [kəˈmɜːʃəlɪ] adv. 商业上地，经济地
beverage [ˈbevərɪdʒ] n. 饮料
matrix [ˈmeɪtrɪks] n. 基质
matrices [ˈmeɪtrɪsiːz] n. 基质（复数）
fluorescence [fləˈresns] n. 荧光，荧光性
scarce [skeəs] adj. 缺乏的，不足的
tandem [ˈtændəm] n. 串联工作组
metabolite [mɪˈtæbəˌlaɪt] n. 代谢物

Expressions and Technical Terms

monomeric intermediate 中间单体
polycarbonate plastics 聚碳酸酯塑料
epoxy resin 环氧树脂
hormone estrogen 荷尔蒙雌激素
human body fluid 人体体液

chromatographic separation step 色谱分离的步骤
mass spectrometry detection 质谱法检测
MS/MS 串级质谱法

Exercises

A. Translate the following into English or Chinese.

1. 双酚 A   2. 基质   3. 替代品   4. 串联工作组

5. monomeric intermediate   6. polycarbonate plastics

7. epoxy resins   8. mass spectrometry detection

B. Decide whether the following statements are true (T) or false (F). Write T for ture and F for false in each blank.

(   ) 1. The exposure of BPA to humans can be through the diet, air, water, and dust.

(   ) 2. BPA-free products have been introduced commercially and the two most common replacements are bisphenol S (BPS) and bisphenol F (BPF).

(   ) 3. In all cases, reliable analytical methods are essential to determine BPA (and its replacements).

(   ) 4. Although the published determinations of BPS and BPF are scarce, methods different to those for BPA should be applicable since the compounds are so structurally different.

(   ) 5. The most sensitive and specific methods use GC and HPLC with MS or MS/MS detection.

C. Translation.

1. Extraction is usually followed by a chromatographic separation step with fluorescence or mass spectrometry detection.

2. The sensitivity and specificity of methods for BPA and its substitutes have improved greatly with the introduction of tandem mass spectrometric methods and ultrahigh-pressure LC.

Appendix

附录

Appendix I
Head and End of Common English Words
英语常用词头和词尾

有许多英语单词是由词的基干部分加上词头或词尾构成的，例如：oxide（氧化物）加上词头 di-（二）成为 dioxide（二氧化物）；work（工作）加词尾 -er（者）成为 worker（工人）；polytetrafluoroethylene（聚四氟乙烯）是由 poly-（聚），tetra-（四），fluoro-（氟），ethylene（乙烯）构成。学习词头词尾对词义理解和单词的记忆很有帮助，现将常用的词头和词尾列表如下。

表 I-1　常用词头

词头	意义	例词
anti-	反抗	anti-imperialism 反帝国主义（imperialism 帝国主义）
bi-	两个	bichloride 二氯化物（chloride 氯化物）
centi-	百分之一	centimeter 厘米（meter 米）
co-	共同,联合	cooperation 合作（operation 作用）
col-		collocate 排列（locate 安排）
com-		composition 组成（position 位置）
con-		confirm 使坚定（firm 坚定）
de-	脱,除	decomposition 分解（composition 组成）
di-	二	dioxide 二氧化物（oxide 氧化物）
tri-	三	tributyl amine 三丁胺（butyl amine 丁胺）
dis-	分开,除去	discover 发现（cover 覆盖物）
en-	使	enlarge 使扩大（large 大）
hydro-	氢或水	hydrometer 液体比重计（meter 计量器,仪表）
im-	不,无	impossible 不可能的（possible 可能的）
ir-		irresistible 不可抗拒的（resistible 可抗拒的）
in-		incorrect 不正确（correct 正确）
inter-	相互,在……之间	interchange 交换（change 变化）
iso-	异	isobutene 异丁烯（butene 丁烯）
kilo-	千	kilogram 千克,公斤（gram 克）
milli-	千分之一	millimeter 毫米（meter 米）
mis-	错误	misunderstand 误解（understand 了解）

（续表）

词头	意义	例词
multi-	多	multi-purpose 多重用途（purpose 目的,用途）
non-	非,不	non-ferrous metal 非铁金属,有色金属（metal 金属）
poly-	多,聚	polymer 聚合物 polyoxide 多氧化物（oxide 氧化物）
pre-	预先	preheat 预热（heat 热）
re-	重复	refit 重新装配（fit 适合,装配）
sur-	在上,胜	surpass 胜过（pass 通过）
tele-	过,远	telephone 电话（phone 声）
trans-	反式	trans-addition 反式加成（addition 加成）
un-	相反,不	unequal 不等的（equal 相等的）

表Ⅰ-2　常用词尾

词类	意义	词尾	例词
构成名词	表示人或物	-er -or -ist	worker 工人 indicator 指示剂 communist 共产主义者
构成名词	表示行为、性质、状态等	-ion -ing -ance -ence -ment -ure -ics -age -ness -ity -y	revolution 革命 teaching 教导 importance 重要（性） difference 差异 movement 运动 pressure 压力 physics 物理学 percentage 百分比 correctness 正确 purity 纯度 difficulty 困难
构成序数词	第……	-th	sixth 第六
构成形容词	具有某种性质	-ic -al -ful	atomic 原子的 experimental 实验的 careful 小心的
构成形容词	具有某种特征	-ous -able -ive -ant -ent	fibrous 纤维（状）的 changeable 可变的 active 积极的,活泼的 important 重要的 different 不同的
构成形容词	不具有某种特征	-less	colourless 无色的
构成副词	表程度、方式、方向	-ly -ward(s)	carefully 小心地 eastward 向东
构成动词	使	-ize -en -fy -ate	criticize 批评 harden 变硬 purify（使）纯化 separate（使）分离

表 I-3　表示数目和数量的常用词头

词头	英文含义	意义	例词
semi- demi- hemi-	half	半	semiconductor, semicircle, demilune, demisemi, hemisphere, hemihedral
uni- mono-	one, single	单，一	uniaxial, unilateral, unicellular, monoacid, monoxide, monohydric
sesqui-	one and a half	倍半，二分之三	sesquiester, iron sesquioxide, sesquicentennial, iron sesquisulfide
bi-, bin-（元音前） di-	two, twice	双，二	biaxial, bilateral, binaural, binocular, dichloride, divalent, dilemma
ambi- amphi-	both	双，两	ambidexter, amphibian, amphicar, amphitheatre
ter- tri-	three, thrice	三	tertiary, tervalent, trisect, trinitrotoluene, trigonometry
tetr(a)- quadri-（辅音前） quadr-（元音前） quadru-（p音前）	four	四	tetroxide, tetrode, tetravalent, quadrilateral, quadrant, quadrupole, quadruple, quadruped
pent(a)- quinqu(e)-	five	五，戊	pentagon, pentatomic, pentahedron, quinquelateral, quinquennial
hex(a)- sex(i)-	six	六，己	hexagon, hexyl alcohol, sexivalent, sexennial, sextuple, sextant

Appendix II
Common Chemical Prefix and Suffix
化学专业英语词汇常用前后缀

表 II-1 常用前后缀

前/后缀	意义	前/后缀	意义
-acetal	缩醛	chloro-	氯代
acetal-	乙酰	chromo-	铬的
acid	酸	cis-	顺式
-al,aldehyde	醛	-cide	除……剂,防……剂
alcohol,-ol	醇	cyclo-	环
alkali-	碱	deca-	十
allyl	丙烯基	deci	分,10^{-1}
alkoxy-	烷氧基	-dine	啶
-amide	酰胺	dodeca-	十二
amino-	氨基的	-ene	烯
-amidine	脒	epoxy-	环氧
-amine	胺	-ester	酯
-ane	烷	-ether	醚
anhydride	酐	ethoxy-	乙氧基
anilino-	苯胺基	ethyl	乙基
aquo-	含水的	ferro-	亚铁
-ase	酶	fluoro-	氟代
-ate	含氧酸的盐、酯	formyl-	甲酰基
-atriene	三烯	-glycol	二醇
-atriyne	三炔	hendeca-	十一
azo-	偶氮	hepta-	七
bis-	双	heptadeca-	十七
-borane	硼烷	hexa-	六
bromo-	溴	hexadeca-	十六
butyl	丁基	homo-	均,同,单
carbonyl	羰基	hydroxyl	羟基
-carboxylic acid	羧酸	hyper-	高,超
centi-	厘,10^{-2}	hypo-	低级的,次

（续表）

前/后缀	意 义	前/后缀	意 义
-ic	酸的,高价金属离子	per-	高,过
-ide	无氧酸盐,酐	petro-	石油
-imine	亚胺	phenol	酚,苯酚
iodo-	碘代	phenyl	苯基
-iridine	丙啶	-philic	亲……的
iso-	异,等,同	pico-	皮,10^{-12}
-ite	亚酸盐	poly-	聚,多
keto-、ketone、-one、oxo-	酮	quadri-	四
-lactone	内酯	quinque-	五
mega-	兆,10^6	semi-	半
meta-	间,偏	septi-	七
methoxy-	甲氧基	sesqui	一个半
methyl	甲基	sexi-	六
micro-	微,10^{-6}	-side	苷
milli-	毫,10^{-3}	sub-	亚
mono-	一,单	sulfa-	磺胺
nano-	纳,10^{-9}	sulfo-	磺酸基
nitro-	硝基	sym-	对称
nitroso-	亚硝基	syn-	同,共
nona-	九	tetrakis-	四个
nonadeca-	十九	ter-	三
octa-	八	tetra-	四
octadeca-	十八	tetradeca-	十四
-oic	酸的	thio-	硫代
ortho-	邻,正,原	*trans*-	反式,超,跨
-ous	亚酸的,低价金属	tri-	三
oxa-	氧杂	trideca-	十三
-oxide	氧化合物	tris-	三个
-oxime	肟	undeca-	十一
oxy-	氧化	uni-	单,一
-oyl halide	酰卤	unsym-	不对称的,偏位
-oyl	酰	-yl	基
para-	对位	-ylene	亚基
penta-	五	-yne	炔
pentadeca-	十五		

表 Ⅱ-2 表示否定意义前缀

前缀	含义及用法	例词
a- an-	在 n. 或 adj. 前，表示缺乏某性质；在元音或 h 字母前	aperiodic, asynchronous, asymmetry, anhydrous, anisotropic, anonymous
anti-	在 n. 前，表示抗……，防……，与……反	antibody, anti-parallel, antifoamer
de-	在 v. 或 n. 前，表示反动作	decolor, decentralize, decrease, descend
dis-	在 n.、v. 或 adj. 前，表示否定或相反，加在否定意义词前，加强否定	disassociate, discharge, disproof, discrete, disannul
mis-	在 n.、v. 或 adj. 前，表示错误	miscalculate, misunderstanding
im-/ in-/il-/ir-	im-用在 b、p、m 之前，表示"不、无、非"	impossible, imbalance, incomparable, illocal, irregular, irrelative, impure
non-	表示"不、无、非"	non-coking coal, non-elastic, nonmetal

表 Ⅱ-3 表示位置或相互关系的前缀

前缀	含义	例词	前缀	含义	例词
ortho-	邻位	ortho-compound, ortho-effect	inter-	相互间	intermolecular, international
meta-	间位	meta-compound, meta-derivative	intra-	在内部	intramolecular, intracrystalline
para-	对位	para-dioxybenzene	iso-	相同,异	isotope, isooctane, isobar
meso-	中位	meso-phase, meso-position	hyper-	超,高	hypertension, hyperchlorate
trans-	反式	*trans*-2-butene, transferase	hypo-	次,低	hypotension, hypochlorite
cis-	顺式	*cis*-1,3-butadiene	sub-	亚,微	sub-atomic, subacid

Appendix III
Common Chemical English Words
化学专业常用词汇

Foundation 基础知识

基础部分

国际纯化学与应用化学联合会	IUPAC	物理分析	physical analysis
分析化学	analytical chemistry	物理化学分析	physicochemical analysis
定性分析	qualitative analysis	仪器分析法	instrumental analysis
定量分析	quantitative analysis	化学计量学	chemometrics

分析数据处理部分

绝对误差	absolute error	平均偏差	average deviation
相对误差	relative error	相对平均偏差	relative average deviation
系统误差	systematic error	标准偏差（标准差）	standard deviation
随机误差	accidental error	相对标准偏差	relative standard deviation
准确度	accuracy	误差传递	propagation of error
精密度	precision	有效数字	significant figure
偏差	deviation	置信水平	confidence level

Quantitative Analysis 定量分析

滴定分析概论

滴定分析法	titrimetric analysis	容量分析法	volumetric analysis
滴定	titration	化学计量点	stoichiometric point

酸碱滴定法

酸碱滴定法	acid-base titration	变色范围	colour change interval
质子平衡式	proton balance equation	混合指示剂	mixed indicator
酸碱指示剂	acid-base indicator	双指示剂滴定法	double indicator titration
指示剂常数	indicator constant		

非水滴定法

非水滴定法	nonaqueous titration	碱性溶剂	basic solvent
质子溶剂	protonic solvent	两性溶剂	amphoteric solvent
酸性溶剂	acid solvent		

配位滴定法

配位滴定法	compleximetry	螯合物	chelate compound
乙二胺四乙酸	ethylene diamine tetraacetic acid (EDTA)	金属铬指示剂	metalchrome indicator

氧化还原滴定法

氧化还原滴定法	oxidation-reduction titration	高锰酸盐滴定法	permanganate titration
碘量法	iodometry	外指示剂	external indicator, outside indicator
溴量法	bromometry	重铬酸盐法	dichromate titration
铈(Ⅳ)量法	cerimetry		

沉淀滴定法

沉淀滴定法	precipitation titration	银量滴定法	argentometric titration
容量沉淀法	precipitation volumetry		

重量分析法

重量分析法	gravimetric analysis	溶剂萃取法	solvent extraction
挥发法	volatilization method	反萃取	reverse extraction
吸入水	water of imbibition	分配系数	partition coefficient
结晶水	water of crystallization	沉淀形式	precipitation form
组成水	essential water	称量形式	weighing form
液-液萃取法	liquid-liquid extraction		

仪器分析部分

物理分析	physical analysis	仪器分析	instrumental analysis
物理化学分析	physicochemical analysis		

电位法

电化学分析法	electrochemical analysis	参比电极	reference electrode
电解分析法	electrolytic analysis	标准氢电极	standard hydrogen electrode
化学电池	chemical cell	一级参比电极	primary reference electrode
原电池	galvanic cell	饱和甘汞电极	saturated calomel electrode
电解池	electrolytic cell	银-氯化银电极	silver silver-chloride electrode
负极	cathode	液接界面	liquid junction boundary
正极	anode	复合pH电极	combination pH electrode
电动势	electromotive force	离子选择电极	ion selective electrode
指示电极	indicating electrode		

光谱分析法概论

普朗克常数	Planck constant	光谱分析法	spectroscopic analysis
光谱	spectrum	质谱图	mass spectrum
原子发射光谱	atomic emission spectroscopy	质谱学	mass spectroscopy

紫外-可见分光光度法

紫外-可见分光光度法		ultraviolet and visible spectrophotometry (UV/Vis)	
红移	red shift	透光率	transmittance
光谱红移	bathochromic shift	吸光度	absorbance
紫移	hypochromic shift	带宽	band width
蓝(紫)移	blue shift	杂散光	stray light
强带	strong band	噪声	noise
弱带	weak band	暗噪声	dark noise
吸收带	absorption band	散粒噪声	shot noise

红外分光光度法

红外线	infrared ray	红外光谱法	infrared spectroscopy
中红外吸收光谱	mid-infrared absorption spectrum	红外分光光度法	infrared spectrophotometry
远红外光谱	far-infrared spectrum	振动模式	mode of vibration
微波谱	microwave spectrum	特征吸收	characteristic absorption

原子吸收分光光度法

原子光谱学	atomic spectroscopy
原子吸收分光光度法	atomic absorption spectrophotometry
原子发射分光光度法	atomic emission spectrophotometry
原子荧光光谱法	atomic fluorescence spectrometry

核磁共振波谱法

核磁共振	nuclear magnetic resonance	核磁共振波谱法	NMR spectroscopy
核磁共振谱	NMR spectrum	化学位移	chemical shift

质谱法

质谱法	mass spectrometry	质谱	mass spectrum
基峰	base peak	分辨率	resolution
质量范围	mass range	灵敏度	sensitivity
气相色谱-质谱法	gas chromatography-mass spectrometry		
高效液相色谱-质谱法	high performance liquid chromatography-mass spectrometry		

色谱分析法

色谱法(层析法)	chromatography	填充柱	packed column
固定相	stationary phase	纸色谱法	paper chromatography
流动相	mobile phase	薄层色谱法	thin layer chromatography

气相色谱法	gas chromatography	离子交换色谱法	ion exchange chromatography
液相色谱法	liquid chromatography	分配系数	distribution coefficient
高效液相色谱法	high performance liquid chromatography		

气相色谱法

气相色谱法	gas chromatography	保留体积	retention volume
保留时间	retention time	死时间	dead time
调整保留时间	adjusted retention time	半峰宽	peak width at half height
峰宽	peak width	等温线	isotherm
灵敏度	sensitivity	漂移	drift(d)
分离度	resolution	检测限	detectability
归一化法	normalization method	外标法	external standard method
理论塔板高度	height equivalent to a theoretical plate		
热导检测器	thermal conductivity detector		
氢火焰离子化检测器	flame ionization detector		
电子捕获检测器	electron capture detector		

高效液相色谱法

正相	normal phase	反相	reversed phase
紫外线探测器	ultraviolet detector	荧光检测器	fluorescence detector
筛分	sieving		

Reagent 化学试剂

优级纯试剂	guarantee reagent(GR)	分析纯(试剂等级)	analytical reagent(AR)
化学纯	chemical pure(CP)	实验室试剂	laboratory reagent(LR)
超纯试剂	extra pure reagent(EP)	特纯	purissimum(Puriss)
超纯	ultra pure(UP)	精制	purify(Purif)
分光纯	ultra violet pure	光谱纯(的)	spectroscopically pure(SP)
显示剂	developer	指示剂	indicator(Ind)
配位指示剂	complexion indicator (Complex ind)	荧光指示剂	fluorescence indicator (Fluor ind)
氧化还原指示剂	redox indicator(Redox ind)	吸附指示剂	adsorption indicator(Adsorb ind)
基准试剂	primary reagent	光谱标准物质	spectrographic standard substance(SSS)
被分析物	analyte	滴定剂	titrant
基准物	primary standard	标准溶液	standard solution

酸碱滴定 acid-base titration

无水碳酸钠	anhydrous sodium carbonate	邻苯二甲酸氢钾	potassium hydrogen phthalate
盐酸	hydrochloric acid	硫酸	sulfuric acid
硝酸	nitric acid	冰乙酸	glacial acetic acid
硼酸	boric acid	氢氧化钾	potassium hydroxide
氢氧化钠	sodium hydroxide	氢氧化钙	calcium hydroxide
硬脂酸	stearic acid	氯化钠	sodium chloride
乙酸铵	ammonium acetate	乙酸钾	potassium acetate
氯化锂	lithium chloride	石蕊试纸	litmus paper
甲基橙	methyl orange	甲基红	methyl red
亚甲基蓝	methylene blue trihydrate	酚酞	phenolphthalein
石蕊	litmus	偶氮紫	azo violet
溴甲酚绿	bromocresol green	溴酚蓝	bromophenol blue
百里酚酞	thymol phthalein	甲酚红	cresol red
结晶紫	crystal violet	百里酚蓝	thymol blue

配位滴定 complexometric titration

氧化锌	zinc oxide	EDTA	ethylene diamine tetraacetic acid
氧化钙	calcium oxide	铬黑 T 指示剂	eriochrome black T indicator
高锰酸钾	potassium permanganate	氨水	ammonium hydroxide
硬度	hardness		

氧化还原滴定 oxidation-reduction titration

硫代硫酸钠	sodium thiosulfate	碘量法	iodometry
碘化钾	potassium iodide	淀粉指示液	starch
碘	iodine	重铬酸钾	potassium dichromate
溴	bromine		

沉淀滴定 precipitation titration

硝酸银	silver nitrate	氯化钠	sodium chloride
硫酸铁铵指示液	ferric ammonium sulfate	铬酸钾指示液	potassium chromate
荧光素	fluorescein	曙红钠	eosin sodium
硫氰酸铵	ammonium sulfocyanate		

其他常用无机试剂

中性氧化铝	aluminum oxide(neutrality)	过氧化铜(粉干状)	copper dioxide powder
镁带	magnesium ribbon	镁粉	magnesium powder
锌片	zinc plate	锌粉	zinc powder

Continued

还原铁粉	iron powder reduced	铁丝	iron wire
金属钠	sodium	金属钾	potassium
铝粉	aluminum powder	过氧化氢	hydrogen peroxide
锂粒	lithium particle		

其他常用有机试剂

甲醇	methanol	正丙醇	n-propanol
无水乙醇	ethanol absolute	葡萄糖	glucose
正丁醇	n-butanol	异丙醇	isopropanol
蔗糖	sucrose	异戊醇	isoamyl alcohol
聚乙二醇	polyethylene glycol	氯仿	chloroform

Common Laboratory Instruments
化学实验室常见仪器

量杯
measuring glass

烧杯
beaker

量筒
graduated/measuring cylinder

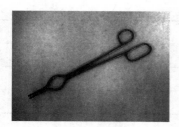

坩埚钳
crucible tongs

坩埚
crucible

试管
test tube

Appendix 附录

试管架
test tube rack

分液漏斗
separating funnel

烧瓶
flask

锥形瓶
conical flask

塞子
stopper

洗瓶
washing bottle

水银温度计
mercury thermometer

试剂瓶
reagent bottle

玻璃棒
glass rod

口（面）罩
mask

碘量瓶
iodine flask

容量瓶（量瓶）
volumetric flask/measuring flask

移液管
pipette

吸量管
graduated pipette

称量瓶
weighing bottle

巴氏吸管
pap straw

天平 balance/scale
台式天平 platform balance

分析天平
analytical balance

洗耳球
rubber suction bulb

研钵
mortar

加热器
heater

滴瓶
dropping bottle

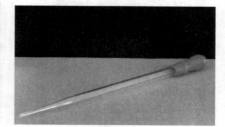
滴管
dropper

升降台
lift（table）

Appendix 附录

铁架台
iron stand

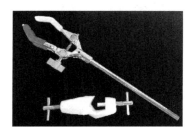
万能夹
extension clamp/universal clip

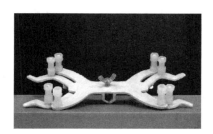
蝴蝶夹
double-burette clamp/butterfly clip

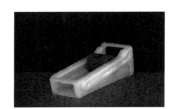
止水夹
flatjaw/pinchcock clamp/water stop clip

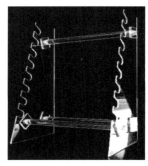
吸移管架
pipette rack

沸石
zeolite

镊子
forceps/tweezer

橡胶管
rubber hose

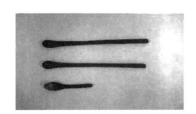
药匙
lab spoon/spatula

pH 试纸
pH paper

滤纸
filter paper

称量纸
weighing paper

擦镜纸
lens cleaning/wiping paper

秒表/停表
stop watch

滴定管
burette

酸式滴定管
flint glass burette with glass stopcock

棕色滴定管（酸式）
brown glass burette with glass stopcock

碱式滴定管
flint glass burette for alkali

（电）磁搅拌器
magnetic stirrer

烘箱
oven

黏度计
viscometer

折射计
refractometer

真空泵
vacuum pump

浴、槽
bath

Appendix 附录

闪点仪
flash point tester

马弗炉
muffle furnace

自动滴定仪
automatic titrator

气相色谱仪
gas chromatograph

红外光谱仪
infrared spectrometer

紫外-可见分光光度计
UV-visible spectrophotometer

高压/效液相色谱仪
high pressure/performance
liquid chromatograph

原子荧光光谱仪
atomic fluorescence
spectrograph

离子色谱仪
ion chromatograph

pH 计
pH meter

原子吸收光谱仪
atomic absorption
spectrometer

电感耦合等离子体原子发射光谱仪
inductively coupled plasma atomic
emission spectrometer

Manipulation 基本操作

分析	analysis	丢弃	discard
氧化	oxidize/oxidation	测定	determine/monitor/test
中和	neutralize	还原	reduce/reduction
滴定	titrate	沉淀	precipitate
滴	drop	溶解	dissolve
样品	sample	加入	add
倾注/慢慢倒入/轻轻倒出	decant	混合	mix
吹吸	blowing-suction	吸入	aspirate
冷却	chill down	通风	ventilate
洗脱	elute	稀释	dilute
搅拌	stir/agitate	蒸馏	distil/distill
脱水	dehydrate	水合,水化	hydrate
过滤	filtrate	燃烧	combustion
研磨	grind	催化剂	catalyst
抽滤	suction filtration	破碎	crush
氢化	hydrogenate	电解	electrolysis
分解	dissolution	合成	synthesis

Appendix IV
Common Chemical Abbreviation and Symbols
化学常见英语缩写与符号

英文缩写和符号	英文全称	中文意思
Å	angstrom unit	埃(10^{-10}m)
ab.	absolute	绝对的
a. c.	alternative current	交流电
addn.	addition	添加
alc.	alcohol	醇
alk.	alkali	碱
amt.	Amount	量
anhydr.	anhydrous	无水的
A. P.	analytically pure	分析纯
app.	apparatus	装置
approx.	approximate	大约
aqu.	aqueous	水的
asym.	asymmetric	不对称的
atm.	atmospheric	大气压
av.	average	平均的
b. p.	boiling point	沸点
℃	Celsirs	摄氏度
c	concentration	浓度
ca.	circa	大约
cal.	calorie	卡(路里)(1cal= 4.1868J)
calc.	calculate	计算
cc	cubic centimetre	立方厘米
cf.	compare	比较
chem.	chemistry	化学
cm	centimetre	厘米
conc.	concentrated	浓缩的
const.	constant	常数
contg.	containing	含有……的

(续表)

英文缩写和符号	英文全称	中文意思
compd.	compound	化合物
C. P.	chemically pure	化学纯
cpd.（compd.）	compound	化合物
cryst.	crystalline	晶体
d.	diameter	直径
d.	density	密度
d-	dextrorotatory	右旋的
D-	dextro-configuration	右旋的
d. c.	direct current	直流
decomp.	decompose	分解
deg.	degree	度
deriv.	derivative	衍生
detn.	determination	测定
dil.	dilute	稀释的
distd.	distilled	蒸馏的
e. g.	for example	例如
elec.	electric	电的
eq.	equation	方程
eqpt.	equipment	设备
equil.	equilibrium	平衡
equiv.	equivalent	等价的
et. al.	and others	以及其他
etc.	et cetera	等等
evap.	evaporation	蒸发
expt.	experimental	实验的
°F	Fahrenheit	华氏度
f.	function	函数
fig.	Figure	图
g	gram	克
g/c. c.	gram(s)per cubic centimetre	克每毫升
h.	height	高度
HP.	horsepower	马力
hr	hour	小时
hyd.	hydrous	水的
i	insoluble	不溶的
i. e.	that is	即
ibid.	in the same place	同前所述

（续表）

英文缩写和符号	英文全称	中文意思
kcal	kilo-calorie	千卡
lab.	laboratory	实验室
liq.	liquid	液体
L. R.	laboratory reagent	实验试剂
manf.	manufacture	制造
max.	maximum	最大的
min.	minimum	最小的
min.	minute	分钟
mixt.	mixture	混合物
mL	milliliter	毫升
mol. wt.	molecular weight	分子量
m. p.	melting point	熔点
org.	organic	有机的
ppm.	parts per million	百万分率
ppt.	precipitated	沉淀的
prep.	prepare	制备
psi	pound(s) per square inch	磅/英寸^2
resp.	respectively	分别地
sec.	second	第二
sec.	second	秒
soln. (sol.)	solution	溶液
solv.	solvent	溶剂
sp. gr.	specific gravity	比重
sq.	square	平方
STP	standard temperature and pressure	标准温度与压力
sub.	sublime	升华
susp.	suspended	悬浮的
T	absolute temperature	热力学温度
t.	time	时间,时刻
t	ton	吨
tech.	technical	技术的
Tech. P.	technically pure	工业纯
temp. (Temp)	temperature	温度
V	volume	体积
V	volt	伏[特]
γ	velocity	（反应）速度
vol.	volume	体积

（续表）

英文缩写和符号	英文全称	中文意思
W	watt(s)	瓦[特]
wt.	weight	重量
Ph	phenyl	苯基
→	yields	生成
=($\approx$)	equal	等于(约等)
(°)	degree	度
π		圆周率
△	heat	加热
%	percent	百分比
μ	micron	微,10^{-6}

Appendix V
The International Phonetic Alphabet
国际音标

单元音			
[i:]	[ɪ]	[e]	[æ]
[ɑ:]	[ɒ]	[ɔ:]	[ʊ]
[u:]	[ʌ]	[ɜ:]	[ə]

双元音			
[eɪ]	[aɪ]	[ɔɪ]	[əʊ]
[aʊ]	[ɪə]	[eə]	[ʊə]

新标准英语国际音标表

辅 音			
[p]	[b]	[t]	[d]
[k]	[g]	[f]	[v]
[s]	[z]	[θ]	[ð]
[ʃ]	[ʒ]	[tʃ]	[dʒ]
[tr]	[dr]	[ts]	[dz]
[m]	[n]	[ŋ]	[h]
[l]	[r]	[w]	[j]

Appendix VI
Periodic Table of the Elements 元素周期表

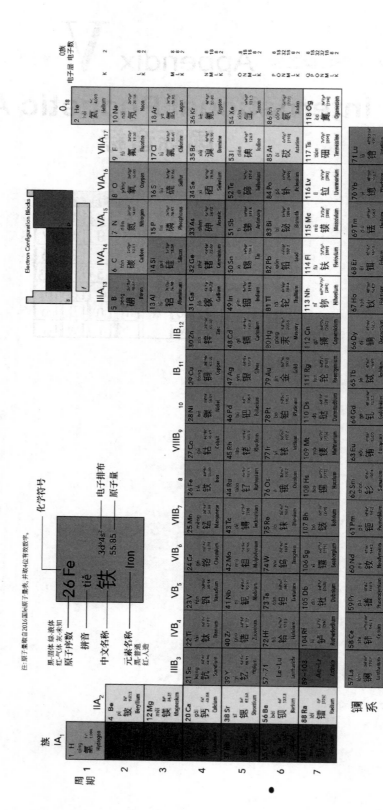

References
参考文献

[1] HAGE D S, CARR, J D. 分析化学和定量分析[M]. 北京：机械工业出版社，2012.
[2] 毕永宏. 化学化工专业英语[M]. 北京：化学工业出版社，1996.
[3] 尹德胜，叶蔚君. 化学化工专业英语[M]. 北京：化学工业出版社，2008.
[4] 李杰，王俊. 应用化学专业英语[M]. 北京：中国石化出版社，2017.
[5] 万有志，王幸宜. 应用化学专业英语[M]. 北京：化学工业出版社，2000.
[6] 符德学. 化学化工专业英语[M]. 北京：化学工业出版社，2011.
[7] Raymond Chang, Kenneth A Goldsby. 基础化学[M]. 北京：科学出版社，2020.
[8] 贾长英，张晓娟. 化工专业英语[M]，北京：中国石化出版社，2018.
[9] 浙江大学外语教研室. 实用科技英语语法[M]. 北京：商务印书馆，1979.
[10] 王慧莉，姜怡. 学术交流英语[M]. 北京：高等教育出版社，2006.
[11] 科学出版社名词室. 英汉化学化工词汇[M]. 北京：科学出版社，2016.
[12] 戴子浠. 有机物国际命名[M]. 北京：中国石化出版社，2004.
[13] Karlheinz Hill. Fat and oils as oleochemical raw materials[J]. *Pure Appl Chem*，2000，72（7）：1255-1264.
[14] 编辑出版委员会. 新牛津英汉双解大词典[M]. 上海：上海外语教育出版社，2007.